"Shawn DuBravac has written a highly intelligent and fair interpretation of the networked future. This is exactly the kind of nuanced and well-informed analysis of our big data world which will make this future more inhabitable. Strongly recommended."

—Andrew Keen, Internet entrepreneur and author
of *The Cult of the Amateur, Digital Vertigo,*
and *The Internet Is Not the Answer*

"Starting from his unique vantage point as Chief Economist for the Consumer Electronics Association, DuBravac takes readers on a non-stop trip to our data-driven, 3D-printed, sensor-rich, autonomous, and automated near future. Better strap in, though. While the destination is inevitable, the road is not without some bumps and sharp curves."

—Larry Downes, *New York Times* bestselling author

"*Digital Destiny* is a welcome antidote to the now all too common stream of ill-informed anti-technology screeds that seek to convince us that the digital revolution is a force for ill. DuBravac sets the record straight, eloquently demonstrating how the emerging data-driven era is a force for progress."

—Robert Atkinson, author of *Innovation Economics*

DIGITAL
DESTINY

DIGITAL DESTINY

HOW THE NEW AGE OF DATA WILL TRANSFORM THE WAY WE WORK, LIVE, AND COMMUNICATE

SHAWN DUBRAVAC, Ph.D.

FOREWORD BY *NEW YORK TIMES* BESTSELLING AUTHOR
GARY SHAPIRO

REGNERY
PUBLISHING
A Salem Communications Company

Library of Congress Cataloging-in-Publication Data

DuBravac, Shawn.
  Digital destiny : how the new age of data will transform the way we work, live, and communicate / Shawn DuBravac.
     pages cm
   ISBN 978-1-62157-373-9 (hardback)
  1. Technology--Social aspects. 2. Technological innovations--Social aspects. I. Title.
   T14.5.D78 2015
   303.48'33--dc23

Published in the United States by
Regnery Publishing
A Salem Communications Company
300 New Jersey Ave NW
Washington, DC 20001
www.Regnery.com

Manufactured in the United States of America

10 9 8 7 6 5 4 3 2 1

Books are available in quantity for promotional or premium use. For information on discounts and terms, please visit our website: www.Regnery.com.

Distributed to the trade by
Perseus Distribution
250 West 57th Street
New York, NY 10107

*To my three digital natives—Nick (eleven), Ryan (nine),
Gavin (seven). May you each secure the destiny you seek.*

CONTENTS

Foreword by Gary Shapiro

Digital Destiny is about the future. It will give you a roadmap to the world we will soon inhabit. It will allow you to make decisions, knowing where remarkable innovation is taking us. It will even give you insight into when and how you should buy products. And it may inspire ideas for you to create a new business.

The legendary ice hockey player Wayne Gretzky said part of his strategy for success was to skate to where the puck is going to be, not where it has been. *Digital Destiny* shows us where the puck of innovation is going to be.

This book will be right in many of its predictions. It will also be wrong in others. Intervening world events and unexpected breakthroughs in technology, costs, and consumer demand can and will affect how and when companies offer products and services and even whether new industries can and will be created.

Even the actions of government can and will affect innovation—a concept I discuss in my 2011 *New York Times* bestseller *The Comeback: How Innovation Will Restore the American Dream*. We see this today as governments fight to protect status quo industries from citizen-to-citizen services such as Airbnb, Lyft, and Uber. In fact, it's why the Consumer Electronics Association (CEA)®, where Shawn and I work, created the Innovation Movement, as we believe fostering innovation should be a key national strategy, to give the next generation a better life.

But despite political uncertainty, I'll bet on Shawn's predictions and vision. I have worked with Shawn for over a dozen years, enjoying scores of his presentations, and marveling at his ability to extrapolate commercial markets from breakthroughs and trends in innovation. Shawn and his colleagues at CEA regularly forecast the next several years of the sales of hundreds of consumer technologies.

Shawn is known for accurate predictions. He told us and the world about the potentially huge market for a portable mid-screen device long before Apple introduced the iPad. He helped accurately forecast the annual sales of HDTV over ten years, long before the product was even in the American lexicon. And he foresaw the popularity of using the real estate of the wrist and body to house new categories of devices.

If it's difficult to predict the near to mid-term future, it's even tougher to write a book about it.

For one thing, this isn't barroom talk or even one of Shawn's compelling keynote speeches. It is a record for history, by which he will be soon judged. In that way, this book is very much like the ambitious and brilliant masterpiece *The Singularity Is Near* by Ray Kurzweil, which meticulously describes progress in several areas of science to present a world decades from now, when new levels of human and computer thought are realized. On the other hand, *Digital Destiny* is accessible to even the least tech-aware reader. By no means simple, it is nevertheless a fun and easy read that turns complex, abstract ideas and principles into compelling, enjoyable prose.

Describing the future in an interesting way is a challenge, because rarely is the real future the stuff of science fiction. We won't have flying cars; there won't be spaceships; and I'm pretty sure we won't have ray guns. As Shawn makes clear, our future is about data—and how human beings will use that data in new, inventive, and life-saving ways. Yet *Digital Destiny* is compelling because it is realistic, useful, and practical.

This book is a rational analysis of where we are going by one of the world's leading experts in trends in technology. In fact, it is colored not only by analysis and experience, but also by Shawn's discussions with and visits to many of the world's leading innovators and technology companies.

Shawn is a participating observer. He thinks big ideas but makes them practical. Eleanor Roosevelt said, "Great minds discuss ideas. Average minds discuss events. Small minds discuss people."

Digital Destiny paints a vision of the intersection of things and ideas on the canvas of the future. Shawn has "emotional intelligence" and knows that every moment we experience we are in one of three places—the past, the present, or the future. This unique perspective allows Shawn to realize the profound implications of small and otherwise ordinary things, such as a sensor in a urinal in Ethiopia. As a technologist, Shawn sees that as evidence of the sensorization of our planet. As an economist, he sees a poor nation like Ethiopia as the beneficiary of the ubiquity of cheap sensors. Put those two together and you have a defining moment—no corner of our world will be untouched by our digital destiny.

Having traveled worldwide with Shawn, I assure you Shawn loves life and is always in the moment. More, he often uses the moment to not only predict but help define the future. Shawn relishes the moment on the slopes skiing or snowboarding, the intense workout in a group class in the weight room where he encourages others through humor and his own effort, or a lengthy run where Shawn sets a mean pace and carries on a conversation.

He experiences the moment of life by filling every minute on every trip with side excursions and gustatory explorations. I remain both thrilled with and angry about a mid-2014 trip experience in Copenhagen where he created his own future and roped me into paying for the best and most expensive meal of my life. More than four hours and twenty-two amazing courses later, we left the restaurant literally the next day, but the bill (with almost no alcohol) topped $1,000. Our corporate rules required me to pay a huge portion of the bill. Thanks, Shawn!

Shawn's ability to squeeze the marrow out of life shapes many of his opinions in this book. Contrary to the current literary rage for novels showcasing a dystopian future, Shawn's vision of the future full of hope and opportunity. He sees new markets for business, nations, and the world. He sees life-saving innovations and great jobs in the future. On the whole, the future we encounter in *Digital Destiny* is a good one. It's one in which I'd be excited to live, and I am happy my children will have that chance. But Shawn's economic mind resists the temptation to paint the future in just one color. He also sees destruction, and chaos, and upheaval. Precisely because the rise of digital data will have such a profound impact on human life, Shawn cannot tuck the reader in and say it's all going to be just fine. It may be just fine, but it'll be a long, rocky road to get there. It's for this reason that *Digital Destiny* again and again reminds the reader that we are at an inflection point in history—one that future generations will look back on and say, "That's when it all changed." It's why the best possible historical analogy for our current moment is the year Gutenberg invented the printing press: 1450. Few appreciated what this German inventor had unleashed on the world—just as few see the big picture of our present moment beyond the cool products and the prospect of driverless cars. Shawn sees beyond the individual products. He sees that what digital data has unleashed on the world is no less monumental than what the printing press unleashed on our forebears. Good or bad, it's going to be one heck of a ride.

Hopefully, you will share my appreciation for Shawn's vision. Innovation is bringing us a great future. Living in the moment does not preclude embracing and planning for the multi-course feast of coming innovation.

> Gary Shapiro
> President and CEO
> Consumer Electronics Association
> *New York Times* bestselling author
> October 14, 2014

Introduction

A s a young boy, I was in a serious car accident. With my siblings and me in the car, my mother fell asleep at the wheel and veered off the road. I distinctly remember the terror that paralyzed me as we plunged into the gully dividing the highway. I couldn't move; I couldn't cry out; I don't think I even breathed. Then it ended as suddenly as it had begun. The car came to an immediate stop. The five of us were dazed but miraculously unharmed.

In 2013, 32,850 people in the United States were not as lucky and died in a motor vehicle accident.[1] In 2012, it was 33,561 people. In 2011, another 32,367. And so on. Sadly, these are considered *good* statistics. As *USA Today* reported following the release of the 2012 numbers, "road deaths are still at the lowest level since 1950."[2] In fact, road deaths have declined in seven of the last eight years.

If we extend our look worldwide, the picture becomes even gloomier. In 2010, according to the World Health Organization, 231,027 people died in India because of car accidents. Although there are no official figures, the WHO estimates that 275,983 people died in China in the same year. All told, the WHO puts the number of worldwide deaths because of car accidents in 2010 at 1.24 million—a number four times the size of the United States population. And this is just an estimate, since the WHO has no way of calculating deaths for a dozen or so countries, such as Libya (population 6.15 million), Somalia (population 10.2 million), and Algeria (population 38.48 million).[3]

Disease and illness aside, if anything else were responsible for the deaths of thirty thousand Americans every year, can you imagine the outrage? Yet aside from some year-end stories announcing a low percentage point drop or rise in the number of road deaths, we take little notice. It's no big secret why. Imperfect human beings operate cars, and unfortunately accidents happen. We accept car fatalities as a tragic fact of modern life.

But what if road fatalities weren't a tragic fact of our technologically infused age? What if instead of hoping for a reduction in the number of road fatalities every year by 2, 3, or 5 percent, we could reduce it by 50, 75, or even 90 percent *forever*? Now extend these numbers across the globe to places like India, China, and elsewhere. Discovering a way to *significantly* reduce car fatalities would be an achievement on par with curing a major disease. It would signal a turning point in history, the end of an era. Some years from now, you would be able to tell your grandchildren that you remember when cars killed thirty thousand Americans every year and more than a million people worldwide. They wouldn't believe it. Those of us who remembered such a time would scarcely believe that we had tolerated it. Imagine it for a minute...more than a million deaths a year.

I'm not talking about what *might be*. I'm talking about what *will be*. The technology to make this a reality is before us today. The introduction of driverless cars will usher in a new paradigm. Driverless cars are our future, our destiny. By removing the human element in car operation we will remove the single factor responsible for nearly all car-related deaths. No more drunk drivers; no more road rage; no more careless lane changes; no more slamming into the car ahead of you because it stopped suddenly. With your concentration no longer focused on the road, you will be able to text, have conversations, watch movies, work, and otherwise make use of countless hours previously spent focused on the task of driving. The book you never have time to read can be finished; the homework your child can't seem to solve without your help, suddenly mastered. The meal you never have time to eat can be enjoyed; the sleep you just can't catch up on, suddenly caught; productivity hours lost to long commutes, suddenly regained.

No longer will parents stay up all night wondering when their child will bring the car back. No longer will you worry about falling asleep at the wheel. No longer will the blind be dependent on others to drive them around.

I don't make this prediction lightly. In fact, my predictions about driverless cars are the easiest ones you will encounter in this book. Driverless cars are not a question of *if,* but *when* and *where*—questions I attempt to answer in the following pages.

To understand why these predictions aren't fanciful, it's important to know my background as chief economist and director of research at the Consumer Electronics Association (CEA), a non-profit trade association representing more than two thousand companies across the consumer tech industry. Every year CEA produces a show that brings together the latest gizmos and gadgets, the smartest minds and most imaginative thinkers across the entire technology ecosystem. In 1967 it started as the Consumer Electronics Show, but now it is now known globally simply as International CES. It's an exciting few days for CEA and the consumer electronics industry. And while the individual gadgets, products, and innovations are the stars of the show, that's not the only thing that excites the mind.

Each year I log over 150,000 miles meeting with companies, speaking with executives, and attending and speaking at industry events. For over ten years, I haven't just seen individual pieces of plastic introduced at CES; I've seen trends develop. I've seen the creation of markets. And what I see today is possibly the single most groundbreaking trend since the advent of the microchip: the digitization of the world around us.

Digital by itself isn't news. By my reckoning, we are well into our second "digital decade." But we have crossed the threshold into a new era of digital technology. Historians like to use the term "revolution" to describe colossal historical moments like this. The First Industrial Revolution and the Second Industrial Revolution were historical periods like the one we are now entering. But they will seem like minor events once the period we are now entering has wholly exerted its force upon us.

We are on the precipice of a new revolution that will utterly transform the way human beings live—and not just human beings in the First World. We already see it today; indeed, we are not bereft of names for our current epoch. We are in the Digital Revolution, the Information Age, or the Computer Age, to cite the most popular.

But the technology and gadgets that gave rise to these titles will one day be seen as mere curiosities compared with what's about to unfold.

To illustrate the point, let's consider a well-known example. In 1998, only 41 percent of households owned a PC, and beyond that the only digital devices households really owned were CD players. At the time, only a very small percentage of households had broadband Internet access. At the dawn of the twenty-first century, we were surprisingly unconnected and operated almost exclusively in the physical, analog realm.

But 1998 was also the year that the first truly digital decade began. In that year, a boutique retailer in San Diego sold the first HDTV, igniting a broad uptake as consumers began to replace their analog devices with digital versions. Soon after came the broad adoption of digital cameras, digital music players (MP3s), digital mobile phones, and a legion of other devices. PC ownership increased over the ensuing years; as of 2015, PCs can be found in more than 90 percent of U.S. households.

Of those who had home Internet access in 2000, only three or four percent had broadband connectivity. Today, just over a decade later, the exact opposite is true. The Pew Research Center's Internet & American Life Project recently reported only three percent of those with a home Internet connection rely on dial-up services.[4]

This is a storyline that has played out across numerous examples. In 2011, 35 percent of Americans owned a smartphone. Just a few short years later, in 2015, nearly 70 percent owned one. In fact, smartphone adoption growth is a rare, almost unprecedented phenomenon. According to MIT researchers, "the only technology that moved as quickly to the U.S. mainstream was television between 1950 and 1953."[5]

Put simply, in a little over ten years the groundswell of digital ownership has reached every corner of our lives. The invention of a single device does not a revolution make. While most of us likely had CD players in the 1990s, the vast majority of our lives were dominated by analog technology, from TV to radio to cameras to phones. Indeed, the CD player augured the closing chapters of the Analog Age, but it wasn't quite over yet. Even today, analog remains present in much of our daily life, much as the typewriter held on well into the 1990s.

But the scale has tipped. This is why I say we're in the "second digital decade," because the first digital decade, as explained above, saw most of us swap analog devices for digital, undergoing a massive digital transformation unlike any transformation we've seen.

I have little doubt that when the history of our era is written, historians won't pinpoint the creation of the PC, the CD player, or the smartphone as the moment human development entered a new digital era. Instead they'll look at when digital overtook analog as the predominant medium through which human technology operates. That time is before us right now.

I am less interested in coining a new term for this new age than in what this new world—an all-digital world—means for all of us in practice. Whether or not we enter the digital age is not a choice anymore. Once man invented the sail, he could not go back to just using the oar. Once man invented the steam engine, he could not pretend that horses were the only way to travel. And once man discovered atomic energy, he had to come to grips with the benefits and risks of his new innovation—from power plants to bombs. Digital technology resides in the same realm as these transformational technologies, and we cannot undo what has been done. This is our digital destiny. This is not what *might* happen if we choose this road over another. This is what *will happen* regardless of which road we take.

Much as the sail allowed man to harness the wind, digital technologies allow us to harness the power of data in a way we never imagined. Like the wind, the data has always been there, surrounding us, but most of it was useless because we couldn't capture it in a systematic way. Until today.

In 2009, the British technologist Kevin Ashton, who coined the term "the Internet of Things," wrote,

> If we had computers that knew everything there was to know about things—using data they gathered without any help from us—we would be able to track and count everything, and greatly reduce waste, loss and cost. We would know when things needed replacing, repairing or recalling, and whether they were fresh or past their best.
>
> We need to empower computers with their own means of gathering information, so they can see, hear and smell the world for themselves, in all its random glory. RFID and sensor technology enable computers to observe, identify and understand the world—without the limitations of human-entered data.[6]

That was six years ago, but Ashton hit the nail squarely on the head. Today, as we increasingly digitize everyday objects—from appliances, to surfboards, to autonomous vehicles, to metrics around our own bodies—we're also embedding sensors in thousands of new devices, many of which are connected to the Internet. This allows us to digitally capture information in a way that accelerates its flow to people, services, and devices. Today's computers, devices, and everyday objects are increasingly gathering data on their own, overcoming the limitations of human-entered data.

In 2008 the number of "things" connected to the Internet surpassed the number of people on the planet. Cisco predicts the number of connected things will grow to between 15 and 25 billion by 2015, before exploding to 40 or 50 billion by 2020. Another eye-popping estimate from Cisco: that 50-billion total predicted for 2020 would represent only four percent of the "things" on earth, a far cry from the level of connected objects we may one day realize.

As objects are digitized and add capabilities such as context awareness, processing power, and energy independence, and as more people and new types of information are connected, the Internet of Things grows exponentially. It becomes a network of networks where billions or even trillions of connections create boundless opportunities for businesses, individuals, and countries.

Indeed, as Cisco Chairman and CEO John Chambers said during his keynote address at the 2014 International CES, "the Internet of Everything will be five to ten times more impactful in the next decade than the entire Internet has been to date." In other words: you ain't seen nothing yet.

There is a tremendous amount of experimentation taking place today. Device makers are leveraging mobile devices to connect things to the Internet that were previously unimaginable, such as keys, coffee pots, thermostats, and health and fitness monitors. And as these connected devices get smaller, faster, and more affordable, their market penetration is poised to take off. What was once technically difficult and not commercially viable because of cost and size is quickly becoming both technically and commercially feasible.

It is critical to note that the Internet of Things is about more than just technology; it's about people. Value will not come just from connecting physical things but from successfully routing the data these connected things capture to the right person, at the right time, and on the right device, to enable better decisions. We are increasingly surrounded by billions of connections; providing these connection points with intelligence will, in turn, influence everything we do.

At the very center of all this turmoil is an ancient concept: data. Human beings have been compiling data since the first cuneiform scripts—used to help ancient traders—were written. As technology has advanced, so has our ability to collect, analyze, and use more and more kinds of data. But when compared to other subjects of technological advancements, data has been relatively stable—because for centuries modes of data collection and distribution remained unchanged. The world may have been using better mechanical technology that allowed us to produce more, travel faster, stay warmer, and eat better, but we still saw and captured data in the same way. Until the advent of digital technologies, the number of pure data "advances" was relatively small: Gutenberg's printing press, the telegraph, Morse code, radio, television, and the telephone. These inventions allowed data to be obtained and consumed at a rate never seen before. But for the most part, advances in the acquisition and transmission of data progressed at a snail's pace compared to other mechanical innovations.

Then digitization happened: finally, we had something that could harness the infinite amount and subsequent power of data. Data is everywhere around us. But most of this data goes unnoticed and uncaptured. How quickly your vehicle consumes gas is a simple example of data that historically was only measured manually. In the past it required you to manually divide the number of miles driven by the reading on your fuel gauge, and because the fuel gauge was analog it was an imprecise exercise at best. It was something you did at intervals rather than continuously, and, if you performed the calculation while you were driving, the data was out of date by definition by the time you finished the arithmetic. But through technological advances this data is now measured digitally and continuously, and algorithms can use it to inform you of things such as how many miles you have left until you run out of gas or your average fuel efficiency.

Data surrounds us. It is how fast you are reading this book right now, your heart rate and blood pressure, your commute time yesterday and every day before that, the length of time you brushed each tooth, and, for that matter, the pressure you applied to each tooth. This is all data. Data, in all its forms, is indeed infinite. But until now, this data was unavailable to us. It existed, but we had no way of recording it in a systematic way and making use of any of it. The simple fact is that human beings were unable to capture and enter all the data that is necessary to make fully informed decisions. It was a task beyond our ability.

Until now.

The fuel of the next industrial revolution will not be mechanical inventions, as it has been through all the ages of history. The lifeblood of tomorrow's world will be data, in all its manifestations. By developing machines that can finally capture and make sense of data, we will unlock solutions to problems that have tormented us since the origin of man. How to eliminate road fatalities is but one of these problems. The digital revolution will also allow us to create solutions to problems we never even knew existed.

That's what our Digital Destiny is really all about. It's not just an abundance of cool gadgets and fun toys. It's not just better TV resolution or safer cars. Humanity's future—our destiny—is an increased ability to harness the power of data through digitization.

CHAPTER 1

The Beginning of Our Voyage and the Properties of Data

*"It has been said that data collection is like garbage collection:
before you collect it you should have in mind what you are going
to do with it."*

—Russell Fox, Max Gorbuny, and Robert Hooke in
The Science of Science

If I had to pinpoint the moment when the idea for this book struck me, it was while walking the convention floor at the 2011 International CES in Las Vegas. After any show, tech writers love debating the best new products and technologies. At the 2011 CES they had plenty to choose from. Over four short days, a plethora of new technologies and innovations had been launched. Ford introduced its first all-electric car and Verizon unveiled its 4G LTE network. Samsung showed off its newest SmartTV technologies, and MakerBot displayed some of the earliest 3D printers. We saw the introduction of a profusion of tablet computers laying the groundwork for 2011 to fast become the year of the tablet. There were first-generation PC ultrabooks, extremely lightweight but powerful laptops, and Microsoft's Xbox Kinect had just entered the market on its way to becoming the fastest-selling consumer electronics device on record, according to Guinness World Records.

It wasn't any particular product or technology that turned on the proverbial light bulb for me back in 2011. It was all the technologies and devices at CES coming together. I was reminded of futurist writer Arthur C. Clarke, who is perhaps best known for the third law in his 1961 book *Profiles of the Future*, which states: "Any sufficiently advanced technology is indistinguishable from magic."[1] As I looked around the halls of CES, I saw magic. I had the immediate realization that a trend once novel and groundbreaking was suddenly quite ordinary and unremarkable—and that made it all the more remarkable.

The digital conversion, underway for more than a decade, was nearly complete. There were few analog products on display. What had been revolutionary in the 1990s, when CES was the birthplace of the first digital cameras and digital televisions, had now become commonplace. And these digital products were now colliding at an increasing rate—creating entirely new devices and services. It's not that this thought hadn't occurred to me before. My job is to watch technology trends and interpret their implications. And I had had plenty of pithy, brilliant things to say about the tablet PC craze or the revolutionary 4G network. I could pontificate all day on the economic impacts of mobile devices or the future of television. Taken individually, these products and technologies are each special and deserving of economic study. And that's what usually happens when you find yourself in the middle of a civilization-altering moment: you take things individually. It's the future historians who are supposed to connect the dots, find the paradigm-shifting trends, and explain what it all meant. It's much harder to do that in the present, and explain what all of these things add up to.

Overcome by this rather simple observation—that the world had gone completely digital—I asked myself the question that led to the book you are reading: What does it mean for us—for human beings—when everything is digital? Is it just a curious trend, like a fashion, changing the way things look, but having little real impact on how we live? Certainly it is more than that. Could it be like the invention of the telephone or the television—game-changing products that forever altered the way human beings receive and provide information? We're getting close, but those are only two products. Of course their impact was revolutionary, but also isolated: you could step away from the television or hang up the phone.

My thinking was still too narrow. Remember, I told myself, *everything* will be digital, not just a few products. So, really, we're talking about something on a larger scale…something like the advent of electricity…ah-ha! I finally felt like I was getting to the essence of the change before us. My simple question had begun to open horizons I hadn't considered. The truly immense scope of digital began to unfold. Digital technology wasn't only going to change what we did and how we did it. It was evolving in a way that would completely transform how cultures are structured and redefine societal norms.

When Thomas Edison invented the world's first light bulb, it was an extraordinary moment. But we're talking about just a single light bulb. How would that change civilization? Were people's homes suddenly wired? Did power stations just spring up over night? Could anyone have had any realistic notion that one day *everything* would be electric? Indeed, as my CEO Gary Shapiro noted in his best-selling book *Ninja Innovation: The Ten Killer Strategies of the World's Most*

Successful Businesses, Edison's true revolutionary innovation was the Pearl Street power station in Manhattan, which provided the current needed to power the light bulbs: "The reason the Pearl Street station was so important—and why Edison would have failed had he not created it—is that the electric light bulb couldn't replace the gas lamp as the primary source of lighting until the entire electrical system was created to sustain it. Otherwise, Edison would have been just the guy who had created a cool, but useless, gadget. In other words, inventing the light bulb was not the end for Edison; it was only the beginning."[2]

The first truly digital consumer product—the CD player—debuted at CES in 1981. It was a big moment, but it did not mark the beginning of the Digital Age. One product cannot do that. Computers go back fifty years. But only a select few of us had one in our homes before the 1980s. Even in the early 80s there wasn't much of a home market for computers, aside from a few niche devices that functioned more as large gaming systems. It was not until Apple brought to market one of the first truly consumer-friendly, *cost-efficient* Macintoshes in 1984 that personal computers became a mass market. *That* marked a paradigm shift; that's the story everyone remembers; that's why Steve Jobs would have earned a place in human history had he done nothing else. It was not the invention of the computer; it was the computer's adoption by the masses.

Twenty years ago, no one owned a smartphone and only the richest few had a mobile device. Today, smartphone penetration is roughly 70 percent in the United States, and people check their devices 150 times a day. The invention itself didn't immediately lead to the end of the phone booth. The era of the phone booth ended only when, decades later, *everyone* had a cell phone.

Let's go back even further. Quick: Who invented the automobile? It's perhaps one of the most important inventions in human history, yet most don't know who is actually credited with inventing it. German engineer Karl Benz (as in Mercedes-Benz) is generally regarded as the inventor of the first automobile powered by an internal combustion engine. That was in 1886. Who owned a car in 1886? Or 1896? Or 1906? Almost no one.

You could be forgiven for thinking Henry Ford invented the automobile. He might as well have. It was Ford, not Benz, who brought the automobile to the masses. Before Ford, the car was just a luxurious alternative to the horse and buggy. After Ford, the car was a necessity. As every school child knows, it was Ford's revolutionary manufacturing techniques that made the automobile a mass market product. In 1909, its first year of production, a Ford Model T cost $850 (nearly $23,000 in today's dollars). By the 1920s, its price had dropped to $250 (about $3,000 in today's dollars) and with its affordability improving dramatically, adoption went up drastically. In November 1929, *Popular Science Monthly* reported, "Of the 32,028,500 automobiles in use in the world, 28,551,500, or more than ninety percent, it is stated, were produced by American manufacturers.... There is, according to these figures, one automobile for every sixty-one persons in the world, an average accounted for by the high ratio in the United States of one automobile for every 4.87 persons."[3]

The automobile age didn't begin when Benz fired up the first car in 1886. It began in 1914 when a Ford assembly-line worker could afford the very product he was making on just four months' salary.

It goes without saying that without Benz, Ford might not have made his first Model T, and had Edison not invented the first light bulb, he would have had no reason to build the Pearl Street power station. Rather, my point is that the existence of a product—its creation—does not transform the world. It's only when that product or technology reaches a critical mass of users or adopters that the human experience changes and a new age is born. Indeed there are countless ingenious inventions that might have fundamentally altered our way of life, but for whatever reason failed to reach a mass market. Either they disappeared or they evolved into something that could reach a mass market.

In short, economics plays as large a role in the progression of human technology as design, engineering, and mathematics. In many cases, it is the dominant role. I can't tell you how many products I've marveled at on the show floor of the International CES that their makers, industry insiders, and critics were sure was The Next Big Thing. Maybe they were right; maybe it would have been. But until it's in the hands of the middle class family of four with two working parents, it's just a novelty.

Digitization of data and devices is today at that inflection point. Digital devices are broadly accessible and becoming increasingly widely adopted. Digital's impact on the world will be more consequential than even the impact of electricity. I recognize that is a bold declaration. In the next chapter we'll explore in greater depth the reasons that the impact of our transition to digital will be so pronounced and significant. For now, let me offer some general observations.

1. Unlike other technologies, including electricity and analog devices, digital technology is subject to Moore's Law. Named after Gordon Moore, whose 1965 paper in *Electronics Magazine* explained the phenomenon, Moore's Law states that the number of transistors on integrated circuits doubles every two years. This leads to two major results: first, processing power continually increases and thus technological innovations continually

occur; second, digital devices become very cheap very quickly.

2. Digital devices and technologies have far exceeded the ownership rates of a wide range of things—from the automobile to many major appliances. A quick look at a few select products tells the story: mobile phones, DVD players, digital televisions, computers, and digital cameras have all achieved an 80 percent penetration rate in the United States:

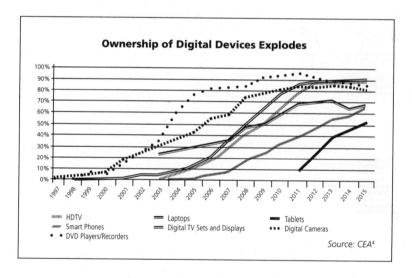

Source: CEA[4]

3. Digital products have found their way into nearly every corner of human society. Much of this book is devoted to exploring both the well-known corners and the ones most of us never imagine. However, digitization will go far beyond what we can see today. Electrical applications start and end with "what needs power," but digital devices start and end with data—and data, we are about to learn, is infinite.

A BRIEF HISTORY OF DATA

To understand digital technology's near-limitless potential we first need to understand its primary function: in its most basic form, digital is simply zeros and ones. But these zeros and ones come together to inform a myriad of things. Digital's ability to process and transfer enormous amounts of data—and I'm referring to amounts that simply dwarf anything that came before it—is what sets it apart from nearly every other human technology. Nothing comes close.

By and large human technology has followed a mechanical progression. First, we invented the wheel, then we made tools, then machines. The Industrial Revolution after all was a revolution in mechanical engineering, first in agriculture, then in manufacturing. We found ways to build better mousetraps. But that progress basically left data untouched.

Indeed, throughout the entire course of human history, revolutions in data transmission and reception have been exceedingly rare. So rare, in fact, that we can recount a brief history here. Let's start with the human brain. The human brain receives surrounding data through its sensory organs: sight, sound, touch, smell, and taste. Our brain also receives internal data signals from the body: pain, hunger, thirst, sleep, and so forth. The history of data can be summarized as man's attempt to recreate the brain's data-processing power.

Take spoken language. What is language but our imperfect attempt to recreate the brain's instantaneous signaling? Without language, we would still know, feel, and desire everything we already do, because that's how our brain operates. It doesn't need spoken language to communicate internally. But we need language to communicate with each other, to describe to one another the messages sent to us by our brain. Language was obviously one of the first revolutions in data transmission and interpretation—allowing human beings to transmit the data in our brains to one another. Although surely there was a time before language, its arrival allowed human beings—as social animals—to live communally, to interact, to support, and to organize more easily. While humans likely did all of these things in some rudimentary form before the formation of languages, linguistic development accelerated all of them. Language precipitated the very beginning of human civilization as we know it.

But language has vast imperfections when it comes to receiving and transmitting data. Its most basic deficiency—and the most significant for our purposes—is that spoken language is transient. It's gone the moment it happens. In the beginning, the only way to sustain data transmitted through language was through human memory. As we all know, this is a suboptimal solution. Nevertheless, early humans adapted in a very simple way: we developed approaches to create better memory. The earliest human societies passed on their myths and history through rote memorization. This lasted well into recorded history, long after written language had already debuted. Many scholars believe that the *Iliad* and the *Odyssey* were originally composed as spoken poems generations before anyone bothered to write them down. Indeed, it was still considered possible in classical Greece and Rome to have one if not both of these extremely long epics memorized.

Nevertheless, the inability of spoken language to sustain itself led a society in Mesopotamia to invent the first written language. In the fourth millennium B.C., the people of Sumer, a region now in modern-day Iraq, created *cuneiform*, believed to be one of the first writing systems, if not the very first. Cuneiform was originally developed as a way to keep track of basic commerce—from a kind of pre-historic bookkeeping that used simple pictograms. Over the centuries it developed into something we'd recognize more properly as a written language, with symbols corresponding to sounds, as in today's alphabet. As primitive as it was, the cuneiform writing system lasted longer than any modern language spoken today, roughly four thousand years.

The era of writing as a mode of data transmission lasted far longer, of course—indeed, in many ways it lasted right up until our modern era. Sure, the speed with which the written word was transmitted quickened because of advances in mechanical technology, but the basic technology for transmitting language would not change until the nineteenth century. That's one heck of a long time for a single innovation to dominate our ability to store and communicate data. That fact helps to underscore the enormity of the data revolution going on today.

But let's not get too far ahead of ourselves. As with spoken language, the written word also suffers from a critical imperfection. Writing had solved the problem of permanence, but added the problem of replication. As a tool for data transmission, it was woefully inadequate because a piece of writing is only useful to the one who has it in his hands. Every piece of writing had to be done by a human being, not to mention a human being knowledgeable enough to know how to do it. In terms of time, this required a tremendous amount of manpower—indeed, too much manpower. Given the choice between educating future scribes to accelerate production of a piece of writing or keeping more hands in the field or in the army, most ancient societies chose the latter.

Although a revolutionary innovation, writing was sparse—and this limited its potential as a solution for the replication of data. Despite the ubiquity of writing in the ancient world, there never was a mass market for reading. It was a luxury reserved for the wealthy elite, much like Benz's automobiles. Data was now permanent, but it was still very much confined.

The problem of the replication and transmission of information troubled human civilization for thousands of years. From the ancient world through the Middle Ages, reading, writing, and learning remained confined to the clergy or the elite; they were simply too expensive for anyone else to enjoy. Books were treasures equal to their weight in the king's gold—that is, until the mid-fifteenth century. Then, everything changed. Data, so long confined, was about to achieve its first escape. It was about to be unleashed.

The dates are speculative, but sometime around 1450 Johannes Gutenberg, a German blacksmith, invented the printing press, complete with movable type and oil-based ink. Just like that, the Middle Ages were over. The printing press revolutionized Europe in a way no other technology has until modern times, and few even come close. The repercussions were more than revolutionary; they were cataclysmic. The Renaissance, the Reformation, the scientific revolution—all of it can be traced to Gutenberg's printing press.

No doubt you're familiar with the history, but consider the innovation of the printing press through the lens of data. Man's millennium-long dilemma of replication had suddenly vanished. In the form of cheap, easily accessible books, data exploded into all corners of Europe. Information, knowledge, literature, the luxuries known only to a select, wealthy few for thousands of years were suddenly everywhere. A single machine erased hundreds of hours of manual work; as a result, the price of books plummeted. A mass market arose. Literacy rates leapt.

What's more, the amount of data increased exponentially. The very term "Renaissance Man" refers to a person who could know everything there was to know. In just a few short decades, that very notion was dead. Now, enabled by the cheap cost of the written word, science equaled what had been achieved by the ancients and then far surpassed it.

The only thing in our lifetime that we can possibly compare to the advent of the printing press is the creation of the Internet. In terms of the eruption of easily accessible, cost-efficient information, the Internet is perhaps the only innovation to have achieved the level of disruption brought on by Gutenberg's press. But even here we speak from a position of privilege, for we live in the post-Gutenberg Age. The Internet merely expanded on a data revolution Gutenberg had begun five hundred years earlier.

For roughly five hundred years, human beings lived in a world Gutenberg had created. There were the communication advancements of the nineteenth century, starting with the telegraph, then the telephone. There were the technological innovations such as the television and radio, which made use of analog signals to broadcast data across distances the printed word could never hope to cross— until the Internet. Indeed, the final problem confining data, which even Gutenberg's printing press could not solve, was distance. Beginning with analog technology, and now with digital, we have solved that problem too. Now no distance is too great for data to travel; entire libraries of data can now be transmitted from anywhere to anywhere nearly instantaneously.

In the next chapter, we'll pick up our brief history of data with the advent of digital. But first let's discuss some of the properties of data that will be paramount in our story.

THE PROPERTIES OF DATA

You may have noticed in our short history lesson that I have granted data volition—that is, a will of its own. I explained how the data revolutions throughout human history had allowed data to do things such as replicate or escape. While the personification of data is a helpful rhetorical device in the writing of its history, I want to take this approach a step further. Data possesses certain characteristics, and to truly see the potential of digital data, it is important to understand how these attributes define its influence.

In his book *What Technology Wants*, Kevin Kelly writes, "we are taught to think of technology first as a pile of hardware and secondly as inert stuff that is wholly dependent on us humans. In this view, technology is only what we make it. Without us, it ceases to be. It does only what we want. But the more I looked at the whole system of technological invention, the more powerful and self-generating I realized it was."[5]

Kelly does a masterful job of making the case for thinking about technology as a living organism. It is my aim to build upon his innovative framework. The leap I ask you to make may not feel natural, but I think it will open your mind and provide an essential paradigm from which to approach data. It will help you see why a book about the 0s and 1s of our world deserves to be a book at all. Join me in anthropomorphizing data.

In describing technology as a living organism, Kelly writes, "The wants of a microscopic single-celled organism are less complex, less demanding, and fewer in number than the wants of you or me, but all organisms share a few fundamental desires: to survive, to grow. All are driven by these 'wants.' The wants of a protozoan are unconscious, unarticulated—more like an urge or a tendency."[6]

Kelly and I do not suggest that data develops wants that are anywhere close to the complexity of the desires of higher beings such as humans. Instead, data resembles the lower living organisms in its faceless and undirected tendencies, its irrepressible momentum hurtling toward everything and nothing. Data has an inevitable future.

Kelly suggests,

technology wants what life wants:

Increasing efficiency

Increasing opportunity

Increasing emergence

Increasing complexity

Increasing diversity

Increasing specialization

Increasing ubiquity

Increasing freedom

Increasing mutualism

Increasing beauty

Increasing sentience

Increasing structure

Increasing evolvability.[7]

If you aren't yet fully convinced of or comfortable with anthropomorphizing data, another way of thinking about these characteristics is not unlike the way we think about water. My dad would always say, "Water seeks its own level," as if water had some choice about where it went. When I lived on the North Shore of Hawaii, my friends and I would often go to Waimea Bay after a heavy rain. During the rainy winter season, the river that cuts through Waimea Valley would flood behind the steep sand dunes of the beach. We would use our hands to dig a small trench slightly below the water level of the dammed river. With very little help from us, the water from the swollen river would start slowly carving its way to the ocean. Of course it was just gravity in action, but it felt as if the water molecules of the river were seeking reunion with those of the ocean and nothing would stand in their way. Steadily and with increasing gusto, the stream would carry sand particles into the sea as it rolled forward—eventually cutting a great chasm in the dune. Ultimately it would create a standing wave where the water from our trench met the ocean (which is why we dug the small channel to begin with). By this point the water was violent, fighting its way into the sea, pushing itself as if obeying a maestro signaling crescendo. And suddenly, after great movement, the water levels would equalize as the river drained, and the wave would dissipate. The whole thing ended as abruptly as it had begun.

We can break down this example and see a number of different states or—if you'll allow—"emotions" emerging: the calmness of the trapped, swollen river, the eagerness of the first trickle of water, the grinding acceleration as it sweeps with it all that lies before it, the joyous reunion when the water hits the sea, the anger and violence as water seemingly turns into force, the stillness when transformation has ended as abruptly as it began.

Of course, water does not feel emotion, but viewing the behavior of water through this very human lens illuminates its principles and power. In other words, assigning water feelings only serves to help us understand and, eventually, use it. As water is placed in different situations—given different opportunities, if you will—it transforms into something seemingly more than its molecular structure would suggest. When acted upon by diverse forces, the water in the example above appears to behave according to its own desires.

The history of data, and more importantly the future of data as realized through its own digital transformation, can be better understood if we see data as possessed of certain properties—characteristics that define not only its movement throughout the ages, but our response to it. As with the water molecules, we exert forces on data; and data, in response to these forces, is becoming seemingly more than its basest structure would suggest. Like water finding its own level, data will find its own level, eventually, because of its properties. I've already mentioned several of those properties, but here is a comprehensive list. Data is:

- **Permanent.** Data seeks permanence. Data exists whether we are around to record it or not, but there has always been a strong human desire to capture data—whether through early storytelling, etchings, writings, or drawings. Across civilizations throughout the history of time data has pushed toward a state of permanence. With the advent of digitization, a growing amount of data can now be stored. While the technology will surely change, it is establishing an environment where data can be stored infinitely.

- **Replicable.** Data wants to replicate. I spoke of this above in terms of data being "unleashed," as if data were locked away like Andy Dufresne in *The Shawshank Redemption*. Through innovations such as Gutenberg's printing press, data has experienced moments of "great escape." In traditional economics, scarcity breeds value, but that is not the case with data, where abundance also creates value. Data replicates because it has value. Replication has long been a defining attribute of data, and one sees a push towards replication wherever one sees data. The arrival of digitalization enables data to replicate exponentially. Nowhere is this more evident than with digital media where a given digital song on Spotify might be "copied" onto a multitude of different playlists.
- **Instantaneous.** Data is immediate and therefore seeks instantaneity. It's only because of the imperfections of human recording and communication mechanisms, whether our minds or our computers, that data comes to us slowly. But data not only wants instantaneous recognition, it wants instantaneous understanding. When data comes into being, when it is first tracked, captured, or copied, it wants to immediately be utilized—to exert force and influence.
- **Efficient.** Data constantly moves toward efficiency. It removes barriers; it closes distances; it destroys the moments between recognition and understanding. Because data wants to be understood, it abhors friction. Data pushes us to invent better, more efficient ways of understanding and disseminating it.

- **Tends toward order.** Data creates chaos, but always on the way towards achieving order. The history of data is a story of order from chaos. Each data revolution, such as the one in which we currently find ourselves, unleashes a maelstrom of information; but in time, humans find ways to put in place a sense of order, a way to find the signal in the noise. Then, when data is further unleashed, the process repeats itself. As a brief example, search engines were invented to instill order in the chaos created by the abundance of the Internet, which unleashed data on a revolutionary scale. But the order of search engines helped unleash a flood of websites so that now even traditional search cannot fully provide order from the chaos digital has unleashed. We need, and we will find, order once again.
- **Moving.** Data is not static. It moves and it seeks movement. Each of the inventions of data transmission in succession—from spoken languages to written records to Gutenberg's press to the Pony Express, telegraphs, telephones, satellites, and the Internet—has accelerated the flow and exchange of information.
- **Infinitely divisible.** Data is infinitely divisible. It can be bundled and unbundled and broken into ever smaller slices. It can be sampled in ever smaller parts.

As you'll see in the pages ahead, many of these properties overlap, and some are more literally applicable to data than others. But the list will serve us well as we dig deeper into the digital revolution unfolding before us. These properties will help us make sense of what we are seeing and what we will see. Significantly, digital technology exacerbates these attributes of data. Understanding these attributes will allow us to see why the digital future before us, in which data will find its freest expression, is not just likely; it is our destiny.

CHAPTER 2

The Seeds of Our Digital Destiny

"In all chaos there is a cosmos, in all disorder a secret order."

—Carl Jung

Where were you?

It's the question that usually begins a discussion about September 11. And if you're of a certain age, everyone has a story to share. What's curious is that it's not really a story about where someone actually was, as in where in the physical universe. Unless one was in New York or Washington, D.C., on that terrible day, the "where" is not really important. Rather, the appropriate and relevant question is: "How did you hear?"

I saw it on the news ...

My mother called me …

I heard it from a co-worker …

You know what almost no one says? *I saw it on the Internet…*Never mind that social media sites like Facebook and Twitter were years away, the act of learning about news—and 9/11 was the biggest breaking news story since the JFK assassination—through a PC or a smartphone was also years away. This is not to say that the news wasn't online in 2001. But the habit of looking to the Internet as a primary news source was still a few years away, as revealed by this 2000 Pew Research Center study: "Fully one-in-three Americans now go online for news at least once a week, compared to 20% in 1998. And 15% say they receive daily reports from the Internet, up from 6% two years ago. At the same time, regular viewership of network news has fallen from 38% to 30% over this period, while local news viewership has fallen from 64% to 56%."[1]

Internet news consumption, while on the rise in 2001, had not reached what we would consider critical mass. Not to jump ahead too quickly, but for comparison, here's Pew on digital news consumption in 2014: "The vast majority of Americans now get news in some digital format. In 2013, 82% of Americans said they got news on a desktop or laptop and 54% said they got news on a mobile device. Beyond that, 35% reported that they get news this way 'frequently' on their desktop or laptop, and 21% on a mobile device (cellphone or tablet)."[2]

It's no wonder that newspaper advertising revenue peaked in 2000 at $49 billion. Ten years later, it was $29 billion—roughly a 40 percent decline.[3] Even though most major newspapers and more than a few regional newspapers had an online presence in 2001, other factors were hindering the use of their sites. For example, U.S. Internet penetration in January of 2001 was just 60 percent. At the time, broadband Internet penetration was only in the single-digits, so it took some effort to "turn on" the Internet. For these reasons, the Internet simply wasn't our go-to source for news. At the time of the September 11 attacks, the weather was the most popular online news feature, according to Pew.[4] On March 14, 2014, the twenty-fifth anniversary of the World Wide Web, 87 percent of U.S. adults used the Internet, "with near-saturation use among those living in households earning $75,000 or more (99%), young adults ages 18–29 (97%), and those with college degrees (97%)."[5]

But perhaps the biggest difference between 2001 and today is in mobile technology. In 2000, 53 percent of U.S. adults owned a cell phone, and no one owned a smartphone, because those devices didn't exist yet. Today, over 90 percent of American adults own a cell phone, with nearly 70 percent owning a smartphone, according to CEA research. Today a high percentage of Americans access the Internet with a device they carry around in their pocket, an impossibility on September 11, 2001.

The September 11 terrorist attacks happened in a world remarkably different from the one we inhabit today. In 2001, nearly all of us received information on the attacks in an analog form. Surely some people learned of the attacks from the Internet, but there are always the exceptions that prove the rule; in the first years of the twenty-first century, Americans were largely living in primarily in the physical, analog world that humans had lived in since the beginning of time.

As the Poynter Institute, a non-profit journalism school, pointed out a year after the attacks: "The horrific events of September 11, 2001, represented the first opportunity for online media—still relatively new—to cover a huge story. While news Web sites were certainly going strong when Timothy McVeigh bombed the federal building in Oklahoma City, even that tragedy pales in comparison to these latest acts of terrorism. This story is of the scope of the attack on Pearl Harbor (some say bigger), and online news is a more mature industry today. Thus, the terrorist attacks on New York and Washington represent the greatest test yet of the newest news medium: the Internet."[6]

How would September 11 have been different in an all-digital world? Almost assuredly you would have heard about the attacks via the Internet, either on your PC or mobile device. We have some proof points that demonstrate how the trend is moving in this direction. In a 2011 opinion piece published by *Forbes* titled "Bin Laden's Death and the Information Revolution," Gary Shapiro noted that Twitter reported, "between 10:45 p.m. and 2:20 a.m. ET users were tweeting at a rate of 3,440 tweets per second" and cited a *Washington Post*/Pew Research poll that found "14% of young people age 18 to 34 heard about bin Laden's death through social networking sites like Twitter and Facebook, a number that rivaled network news (19%) and cable news (17%)."[7]

In a recent survey by the American Press Institute, "half of news consumers who named a breaking news story say they first heard about it from television. About half of those (49 percent of adults) then tried to learn more. Few stuck with television. A majority moved to the Web (using a number of devices) rather than continuing to follow the story on television (59 percent vs. 18 percent)."[8]

Had September 11 happened in an all-digital world, you could have followed the news—perhaps even watched it live—over a broadband network from hundreds of different sources on multiple devices. As painful as it is to imagine, you might have even read live updates from those trapped inside the buildings or on the doomed planes. You would have conversed with friends and family (even victims) over social sites, SMS text, and email. You might have received emergency updates from the government through your mobile phone.

The data on the attacks would have traveled at a speed shocking to our 2001 selves. Not that this would have made the day any easier to bear. In a lot of cases, it would have made events even more chaotic. Today, rumors race across the Internet in a matter of seconds. Imagine the online rumor-mill on September 11. The media, racing to distinguish fact from fiction, would have had quite a mountain of possible leads to pursue. What we experienced on September 11 was horrible, but what we saw and heard was just a fraction of what there was to see and hear. What we don't have are the hundreds of thousands of images, the hour and hours of video, the instantaneous recording of history in the making, which we would have today. The data was there but technological innovation wasn't there yet; we simply didn't have the ability to record, replicate, and transmit it. Now we do—or, at least, our capacity to do so has increased a thousand times over.

The evolution from an analog to an all-digital world doesn't happen overnight, but these changes are symptoms of a future defined by digital.

THE RISE OF DIGITAL

We'll return to the contemplation of a digital September 11 at the end of this chapter, but now let's resume our brief history of data. We left off as we were discussing how Gutenberg's printing press represented data's greatest advance in thousands of years. Indeed, in many ways, we still live in a Gutenberg world, where the ability to replicate data easily (and cheaply) continues unabated. Just sixty years ago, Ray

Bradbury published *Fahrenheit 451*, a dystopian novel in which a totalitarian government keeps order through the destruction of books. Inherent to the plot of the novel is the idea that books are more than a collection of pages with words printed on them—they're knowledge, information, data. If some fictional government entity wished to control the masses, so Bradbury reasoned, it would have first to control the masses' access to data. Books, Gutenberg's children, would be too dangerous to exist. They were *permanent* repositories of data and where one book existed, there was potential for many more to be replicated. Not to give away the ending, but the heroes of the story overcome the data destruction through a prehistoric human tool of data transmission, long since dormant: memorization.

Even in Bradbury's fictional world, the oppressive government had a difficult time destroying every book in existence. The power of Gutenberg's printing press proved itself once again. In today's world, where digital reigns, Bradbury's story would be near impossible to believe. The idea that any government could erase all human knowledge and keep order through data suppression is hard to swallow in the age of the Internet, high-speed broadband, and mobile technology. Governments that have undertaken even the most limited censorship efforts have almost always been frustrated. Which is not to say it's impossible: a quick look at North Korea proves that censoring knowledge across an entire country can happen in a digital world. But it requires keeping the entire country completely isolated from the rest of the world.

Which brings us back to our history course: How did we get here? It would be far beyond the scope of this book to recount every innovation, from the printing press to today, leading to digital data. We'll have to make do with just the highlights, keeping a discerning eye on the characteristics of data we explored in the previous chapter.

Some of the earliest symbolic systems, the precursors of written language, use a limited system of discrete symbols. These systems can't be equated with digital, nor can they be called the beginning of the history of digital, but they do provide the basic building blocks of digital. Take for example an abacus; the counting device invented some three thousand years ago is a system in which the beads only represent numbers while in an "up" or "down" position. Morse code, Braille, and modern maritime signal flags are all contemporary examples of other precursors of digital systems.

Let's now move the clock forward a bit to the invention of binary code, which is the elemental language for all computers and other digital devices. We find ourselves in Germany once more, where a mathematician and philosopher by the name of Gottfried Wilhelm von Leibniz (1646–1716) has "invented" the binary numeral system, which uses just two symbols (0 and 1) to represent all values. Leibniz also appeared to realize what he had discovered. In 1671, he used his binary system to create the world's first mechanical calculator, known as the "Stepped Reckoner." Although the early prototype was temperamental, the machine could perform the four basic arithmetic operations: addition, subtraction, multiplication and division.

But this wasn't the end of Leibniz's innovative foresight. He also imagined a machine that used marbles to represent binary numbers. If you replace the marbles with modern technology, Leibniz basically invented the modern electronic digital computer. But the world would have to wait some 300 years before the technology caught up with the human imagination.

Nearly two hundred years after Leibniz's "Stepped Reckoner" wowed the Royal Society, an English mathematician by the name of George Boole found himself tinkering with his German predecessor's binary system. In 1847, Boole published a paper, "The Mathematical

Analysis of Logic," that featured an algebraic system of logic, based on the binary system. Boole would expand on this work throughout his career, in the process creating Boolean algebra.

Now flash forward another hundred years, when the scene shifts to the New World. A young MIT student by the name of Claude Shannon publishes a landmark paper, "A Symbolic Analysis of Relay and Switching Circuits," which applies Boolean algebra to electrical circuits. The rather dreary title of the paper masks the revolutionary theories unveiled within. One admirer, writing forty years later, called Shannon's paper "possibly the most important, and also the most famous, master's thesis of the century."[9]

Shannon's work was embraced by the scientific community, and events began to accelerate. After reading Shannon's paper, a researcher at Bell Labs named George Stibitz created a basic, yet revolutionary, digital calculator. Using binary addition, the machine, named the Model K (after kitchen, where Stibitz had assembled it) jump-started a new program at Bell Labs. With Stibitz in charge, the program developed the 1940 Complex Number Calculator, which, as its name implies, could work with complex numbers. In a demonstration, Stibitz used a teletype to send commands from a lecture hall at Dartmouth College in Hanover, New Hampshire, to the electromechanical computer in New York over telephone lines. It was kind of a big deal, as a plaque commemorating the event at McNutt Hall at Dartmouth College, reads: "In this building on September 9, 1940, George Robert Stibitz, then a mathematician with Bell Telephone Laboratories, first demonstrated the remote operation of an electrical digital computer. Stibitz, who conceived the electrical digital computer in 1937 at Bell Labs, described his invention of the 'complex number calculator' at a meeting of the Mathematical

Association of America held here. Members of the audience transmitted problems to the computer at Bell Labs in New York City, and in seconds received solutions transmitted from the computer to a teletypewriter in this hall."

A shorter version might read, "Welcome to the Computer Age."

Here is where our story explodes in a thousand different directions, which is why it's probably best to organize some of the major inflection points beside the years in which they occurred.

1947: Scientists at Bell Laboratories invent the first transistor, the key component of nearly every electronic device. The transistor is relatively cheap to produce and much more reliable than the vacuum tubes in use up to this point. Stop and consider the magnitude of this single inflection point. For all the revolutionary value of printing presses, they were expensive, complex machines to operate. You didn't simply have a printing press in your basement. While books were cheap, the machinery needed to produce them was beyond the means of the average consumer. That the transistor, such an important component of all future electronics, was so cost-efficient helped speed the digital revolution that was about to unfold.

1958: Jack Kilby, a scientist working at Texas Instruments, demonstrates the first integrated circuit, which is nothing more than a set of electronic circuits, or transistors, on a single plate. We now commonly refer to this innovation as a microchip.

1964: Gordon E. Moore, co-founder of Intel Corporation, publishes a paper in *Electronics Magazine* with the delightful name, "Cramming More Components onto Integrated Circuits." The paper records Moore's observation that since the invention of the integrated circuit, the number of components on it had doubled every year. Moore concludes, "The complexity for minimum component costs has increased at a rate of roughly a factor of two per year. Certainly over the short term this rate can be expected to continue, if not to increase. Over the longer term, the rate of increase is a bit more uncertain, although there is no reason to believe it will not remain nearly constant for at least 10 years. That means by 1975, the number of components per integrated circuit for minimum cost will be 65,000. I believe that such a large circuit can be built on a single wafer."[10]

With some slight modifications, Moore's observation has proved correct. While mathematicians use the term "law" extremely sparingly, the impact and remarkable longevity of Moore's findings have earned it the moniker "Moore's Law." A graphical representation shows Moore's Law in effect since 1971, when the first commercially available microprocessor came to market.

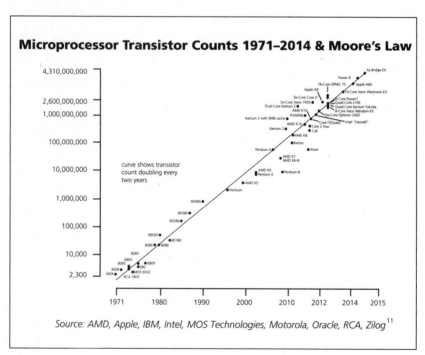

Microprocessor Transistor Counts 1971–2014 & Moore's Law

Source: AMD, Apple, IBM, Intel, MOS Technologies, Motorola, Oracle, RCA, Zilog[11]

While Moore's Law predicts exponential growth for the power of microchips well into the future, it also means digital products become less expensive more quickly than their analog alternatives.

1969: Researchers at the Defense Department's Advanced Research Projects Agency (ARPA) build a computer network named ARPANet, a precursor to the Internet. Designed to connect scientists and researchers with each other, ARPANet first connects one computer in Utah with three in California. The first message sent crashes the system before the first word is finished. The government will later allow other universities and research centers to connect to the system, and by 1975 the network has 57 IMPs (or routers).

1973: Bob Metcalfe, with the help of David Boggs and a few other colleagues, designs and debugs the first working Ethernet prototype while working at Xerox's Palo Alto Research Center (PARC). The first prototype will connect more than a hundred workstation computers on a 1 kilometer cable. Today, every computer connects to the Internet via Ethernet or a derivative.

1975: Steve Sasson, an electrical engineer at the Eastman Kodak Company, invents the world's first digital camera. The ugly thing weighs eight pounds and has only 0.01 megapixels. It stores about thirty images on an analog cassette tape, the only storage device available to Sasson. And the poor picture quality means Sasson's camera will never be a consumer product. When asked when the quality might be "good enough," Sasson predicts between ten and fifteen years, using Moore's Law to estimate when the number of pixels per picture would hit about 2 million.[12]

1981: Sony introduces the first compact disc player for the consumer market at the International CES, the world's largest technology trade event. Eighteen months later, on October 1, 1982, Sony launches the CDP-101. More than a hundred years after Thomas Edison invents the phonograph and thirty-three years after LP records are introduced, CDs transformed the way we listen to music. Digital audio is introduced to the masses. The CD player is the first widely adopted digital product, but it will take nearly twenty more years for digital devices to dominate the consumer electronics marketplace.

1984: Apple releases the Macintosh on January 24 at the reasonable price of $2,495 ($5,595, in 2015 dollars). It isn't as cheap as a typewriter, but the sale of such an advanced product at a relatively low price parallels what Henry Ford did with the Model T seventy years earlier. This marks the culmination of years of work to bring to market affordable mass-produced personal computers that started with the Tandy/Radio Shack TRS-80 in 1977 and continued with introductions from companies like Sinclair, IBM, Compaq and Osborne. Computers leave the confines of the elite, and the era of personal computing begins with strong sales. A contemporary review in the *New York Times*, while lauding the machine, also airs some gripes:

> …MacWrite now has a limited file length of 9 to 10 pages. It is like having a filing cabinet that will hold only folders of the same capacity. If you need to prepare a 20-page report, you will have to separate it into two sections. That is not too convenient. Neither is printing the report.
>
> AGAIN, because of the graphics orientation and the hundreds of thousands of dots required to print each page, the process is much slower than one would like. I was surprised to discover that in order to print out a document on the required dot-matrix printer in anything but rough

form, I first had to have the computer transfer it to the disk and then have the printer take the information from the disk. Since one cannot enter new data while that is going on, there is a long wait.[13]

Although clunky and not terribly efficient, the Mac suddenly turns anyone's office into a Gutenberg printing press. Forget the big industrial monsters sitting in warehouses; now a page can be printed from your personal computer.

1985: Nintendo releases the NES game console, a runaway success. Key to our discussion here, however, is an accessory to the NES called the Nintendo Zapper, or light gun. It uses light sensors to simulate "shots" in a game. We'll be discussing sensors in greater depth in chapter four.

1989: Tim Berners-Lee, an English computer scientist at CERN (European Organization for Nuclear Research), drafts a proposal that essentially describes the World Wide Web. As Berners-Lee later explains, "Creating the web was really an act of desperation, because the situation without it was very difficult when I was working at CERN later. Most of the technology involved in the web, like the hypertext, like the Internet, multi-font text objects, had all been designed already. I just had to put them together. It was a step of generalizing, going to a higher level of abstraction, thinking about all the documentation systems out there as being possibly part of a larger imaginary documentation system."[14]

"An act of desperation" is an interesting way to put it. Here's another way: While working at CERN, Berners-Lee and his colleagues were confronted by a problem of chaos—too much information, not enough efficient organization. Obviously, they want to create order from disorder. Human beings want to create structure. But recall the characteristics of data that I proposed in chapter one. Berners-Lee pushed data towards organization and efficiency. The characteristics I see in data suggest it naturally wants to flow that way too. This "act of desperation" was Berners-Lee pushing the data and the data pushing Berners-Lee towards order, resulting in the World Wide Web.

You can still visit the world's very first web address: http://info.cern.ch/hypertext/WWW/TheProject.html

Absurdly simple by today's standards, the website describes the "the project" of the Web, or W3. Click through a couple of links, and you happen upon this summary:

> The WWW project merges the techniques of information retrieval and hypertext to make an easy but powerful global information system.
>
> The project is based on the philosophy that much academic information should be freely available to anyone. It aims to allow information sharing within internationally dispersed teams, and the dissemination of information by support groups. Originally aimed at the High Energy Physics community, it has spread to other areas and attracted much interest in user support, resource discovery and collaborative work areas.[15]

Connecting the disparate data centers in an "easy but powerful" way made it possible for nearly anyone to access the information. The World Wide Web as originally intended was nothing more than an elaborate system for ordering data. Without it, data was siloed, cut off from meaningful analysis from anyone beyond those physically present to access it. It might not sound like chaos to us, but to Berners-Lee and his colleagues, both at CERN and around the globe, it certainly was.

In the same year, Fuji sells the first consumer handheld digital camera, the DS-X, fourteen years after Sasson predicted it would be possible. At $20,000, the camera is hardly priced for the masses, but Moore's Law would take care of that in relatively short order.

1998: On the morning of August 6, Bruce Colby becomes the first person to buy a high definition digital television. Colby pays $5,499 for that first DTV—a fifty-six-inch Panasonic television—at Dow Stereo/Video, then a specialty retailer of audio and video products in San Diego, California.[16]

1999: On June 1, Sean Parker and Shawn Fanning release the first version of the file sharing service known as Napster. The site allows users to share digital files, usually audio files or songs, encoded in MP3 format. Although file sharing sites have been around since the mid-90s and the MP3 format went back further than that, it is Napster that creates an avalanche of consumer interest. Suddenly, digital files aren't just for the geeks. College students, lured by free music and ease of use, flock to Napster—a blessing and curse for the two founders.

THE CRITICAL CONVERGENCE

The CD player, digital television, digital camera, and digital MP3 player were just the beginning. By the end of the century, digital products flooded consumer shelves, slowly but surely replacing their old analog counterparts in all the little corners of our lives. Over the next decade of the new century, not only did ownership rates increase and broaden, so did ownership density.

As we entered the early "aughts," we began to replace our analog cameras with their digital alternatives. We did the same with our televisions. In 2001, Apple introduced the first iPod, sounding the death knell for the analog music industry. The age of digital devices was upon us, and what ensued was a decade of mass adoption.

In almost all cases, the digitization of devices increased ownership across the board—putting more technology in the hands of users. Moore's Law allowed consumers not only to replace what they had, but also to buy more. Households that historically lacked access to certain technology products because of cost or other factors suddenly found themselves owning not just one digital product, but myriad digital products and in many instances owning multiple digital products within the same category.

In the history we just explored, I focused on three particular trends: computing, the Internet, and digital consumer products. The convergence of these three trends at the end of the century marked the beginning of a new age for humanity. All of these trends took root because of simple economic principles. None of these technological successes would have happened if the device components had not grown less expensive as they grew more powerful. Because of Moore's Law, digital component prices fall faster than analog prices. For example, the quality-adjusted prices of televisions historically fell one to two percent a year until the advent of digital. Today television prices fall one to two percent a month—an annual compound rate of 12 to 14 percent a year. This happened as the technology shifted from tube-based televisions to digital televisions.

At the same time, important shifts in manufacturing took place. Historically, consumer electronics manufacturers owned the entirety of their manufacturing facilities—an efficient if costly strategy. Today, companies like Apple own no manufacturing facilities, while other consumer electronics manufacturers have entered into joint ventures. As the consumer electronics industry increased in popularity, manufacturing consolidated, producing efficiencies and creating economies of scale that improved yields.

Where this matters for us is in the supply chain. The deflationary pressures of digital technology work through the entire supply chain. For instance, semiconductor prices fall by roughly 50 percent every two years, a trend that impacts multiple layers of the supply chain. It starts with the chip manufacturing, continues through the component and device manufacturing and on through distributors and retailers until ultimately the consumer is the beneficiary. No one level of the supply chain—nor even the supply chain in its entirety—is able to capture these price declines for their own benefit and profitability. We have now enjoyed decades upon decades of technological improvement and price declines. We are now at the steep part of the deflationary curve. The price declines enjoyed today are based on decades of price declines, and the power of compounding forces is about to catapult us into an entirely new realm.

In *The Age of Spiritual Machines: When Computers Exceed Human Intelligence*, Ray Kurzweil recounts a story about the invention of chess.[17] It goes something like this: the Emperor at the time was so pleased with the game that he offered the inventor anything in his kingdom as a prize. The inventor requested simply that a single grain of rice be placed on the first square of the chessboard and that it be doubled on each subsequent square. As the story goes, the Emperor was happy to oblige what seemed like such a simple request. That was of course, until he began to see the ramifications. One grain became two grains. Two grains become four grains. At the end of the first row, the Emperor has paid out a total of 255 grains of rice. The grains of rice grow as the power of exponential growth works down the chessboard one square at a time. The second row costs the

Emperor 65,280 grains of rice while the third row costs him over 16.7 million grains, and the fourth over 4.2 billion grains of rice.

Halfway through the chessboard, the Emperor begins to realize he is on the wrong side of the trade. At that point, the Emperor has the inventor beheaded.

Had the story been allowed to continue, we would have seen that the quantity of rice really starts to get big in the second half of the chessboard. While the first half of the chessboard produces almost 4.3 billion grains of rice, the second half of the chessboard, thanks to exponential growth, grows from billions to trillions to quadrillions to quintillions—over 18.4 quintillion, to be more exact. Eric Brynjolfsson and Andrew McAfee recount this same story in *The Second Machine Age: Work, Progress, and Prosperity in a Time of Brilliant Technologies*. Brynjolfsson, McAfee, and Kurzweil all suggest that today we are entering the back half of the chessboard, where the frenzy of growth really kicks in.

But here's one fact they fail to point out. Each new square—each new doubling—is larger than the sum of all the squares before it. Stop and think about that for a minute, and apply this principle to computing today. Roughly every two years computer processing power doubles, and the newly achieved level is greater than the sum of all the exponential growth that existed before it.

Exponential growth creates really big numbers. This is bad if you are an inventor aiming to trick the emperor into a large payout, but great if you are the user of devices with computational power subject to exponential growth pushing us to an entirely new plane. As I pointed out above, no level of the supply chain has been able to capture this exponential growth (or, conversely, the exponential cost declines) for themselves. No one has been able to cut costs in half without pushing all of the savings down the supply chain. Ultimately, the savings fall all the way through every level of the supply chain to the consumer. We are the recipients of every single grain of rice, and what we receive now is larger than the sum of everything before it. Looking ahead, what comes next is going to be larger than what we have today and larger than everything before it. Our Digital Destiny lies before us and it is looking very, very big.

Three important elements—ubiquitous computing, Internet access, and the proliferation of digital consumer products—came together during the first years of the new century. These pieces laid the foundation for our Digital Destiny.

Now in the middle of the second digital decade, the inflection points of this transition surround us. Take music, for example. For the first time, digital music downloads now outsell all other forms of music media, combined. Many digital books now outsell their printed counterparts. Print media such as magazines and dailies now rely on digital dissemination as more and more people consume their news on digital devices. At the dawn of the second digital decade, we are becoming an all-digital society. In fact, according to estimates from Cisco, the number of mobile-connected devices now exceeds the world's population.

Of course, much has been omitted from our brief jaunt through the history of data—not only because you already know it, having lived through it, but also because at this point our story requires another shift in our focus. We'll finish the history of data in the next chapter, where we'll see just what the convergence of computing, the Internet, and digital products has unleashed upon us.

THE FIRST DIGITAL MANHUNT

At 2:49 p.m. on April 15, 2013, two pressure cooker bombs exploded near the finish line of the Boston Marathon. In the aftermath of the attack, three people were dead, with 249 more injured. In scale, the Boston Bombings do not come near the magnitude of September 11. Nevertheless, what happened after the bombs went off can elucidate the question I posed at the beginning of this chapter: What if September 11 had happened in our all-digital world? We can glimpse what might have been through the lens of the Boston tragedy.

Almost immediately after the explosions, social media experienced an explosion of its own. At 2:50 p.m., an eyewitness uploaded a photo of the bombing to Twitter. At 2:51 p.m., another eyewitness tweeted, "An explosion just went off in downtown Boston. Spectators fleeing the #bostonmarathon course." At 2:52 p.m., a local Fox Sports affiliate tweeted, "BREAKING: Per our man on the ground at the Boston Marathon, @tooblackdogs, there was an explosion. More to follow."[18] Television coverage followed quickly, but the first news flashes were pushed out over digital networks.

According to an account in *National Geographic*, the *Boston Globe* temporarily converted its homepage to a live blog that pulled in tweets from Boston authorities, news outlets and citizens.[19] Even the authorities were monitoring the social sites: "Authorities have recognized that one of the first places people go in events like this is to social media, to see what the crowd is saying about what to do next," said Bill Braniff, Executive Director of the National Consortium for the Study of Terrorism and Response to Terrorism.

> And today authorities went to Twitter and directed them to traditional media environments where authorities can present a clear calm picture of what to do next.
>
> We know from crisis communication research that people typically search for corroborating information before they take a corrective action—their TV tells them

there's a tornado brewing and they talk to relatives and neighbors. And now they look at Twitter.

Indeed, #BostonMarathon quickly became a trending hashtag on Twitter. By the end of the week, video footage of the blasts had been tweeted forty thousand times. The Boston Police Department Twitter account grew by more than 273,000 followers, a 500 percent increase. Pew would later report that a quarter of Americans learned about the explosions and the hunt for the bombers from Facebook and Twitter. Among eighteen-to-twenty-nine-year-olds, more than half (56%) got bombing-related news through social networking sites.[20]

As images and video of the explosions and aftermath continued to flood the Internet, law enforcement officials began to pore over the evidence. As the *Washington Post* reported, "The goal was to construct a timeline of images, following possible suspects as they moved along the sidewalks, building a narrative out of a random jumble of pictures from thousands of different phones and cameras."[21] For the first time in history, a physical crime scene was more heavily dissected in the digital realm.

Almost as soon as the bombs exploded, the rumors spread—one of the downsides of disaster in the digital world is the acceleration of data in all of its myriad forms including the spread of rumors and unsubstantiated information. Some said there were four bombs instead of two. It was al Qaeda. No, it was right-wing extremists. No, it was left-wing extremists. And then people began pointing fingers, identifying this person or that person in fuzzy images of the mayhem as the real culprits. An Indian-American student who had been missing for several weeks was falsely identified as one of the bombers.

As the *Washington Post* reported, "On an investigative forum of Reddit.com, since removed from the site, users compiled thousands of photos, studied them for suspicious backpacks and sent their favorite theories spinning out into the wider Internet."

"Find people carrying black bags," wrote the Reddit forum's unnamed moderator. "If they look suspicious, then post them. Then people will try and follow their movements using all the images."[22]

The *New York Post*, falling victim to the frenzy, published an image on its front page of two unidentified men under the headline "Bag Men." The image, it was later learned, was not of the suspects. On April 18, the FBI released to the public an image taken by an eyewitness of the still-unidentified suspects, asking for help. The FBI wanted to put an end to the random speculation swarming around the Web.

In time, the true suspects were identified and eventually run to ground. Again, thousands watched the bloody end unfold over the Internet, as police zeroed in on the two suspects. After it was all over, one was dead; the other, hurt and bloody, was captured after seeking refuge in a parked motor boat.

If we magnify those first few days after the Boston Bombing, we might have a good idea what September 11 would have been like in our digital world. As in Boston, it would have been a story of extreme data capture, replication, and dissemination. It would have given us a far clearer, far more accurate view of the tragedy than what we do have. But it also would have been far more chaotic, perhaps even far scarier than it was. Clarity, after all, doesn't always lead to comfort. And before you can find order, you have to go through the chaos.

CHAPTER 3

When Data Is Digitized

"The current state of knowledge can be summarized thus: in the beginning there was nothing, which exploded."

—Terry Pratchett

On May 20, 2013, Edward Snowden arrived in Hong Kong carrying four laptops. Twelve days later, he met with two journalists for the *Guardian* (UK) newspaper, Glenn Greenwald and Ewen MacAskill, at Hong Kong's Kowloon hotel. Snowden identified himself to the two journalists, who until then had never met their source face to face, using a Rubik's Cube.[1] The little detail of the Rubik's Cube caught my eye, and I want to come back to that in a moment. But first let's take a look at Edward Snowden through the digital lens.

Although the real numbers are classified, it is believed that Snowden took around 1.5 million documents from U.S., British, and Australian intelligence agencies.[2] While we may never know the total number of documents with certainty, imagine for a moment the sheer magnitude of even a fraction of 1.5 million documents. In the analog world, in which we have lived until now, it is simply unfathomable. I estimate that 1.5 million documents would fill a tractor trailer and maybe more. But in our newly found digital world they are part of your carry-on luggage—as was the case with Snowden.

What Snowden did in a digital age would have been almost impossible to do in the analog world. Stealing that sheer volume of physical documents would have required the coordinated efforts of many. But in addition, many of the documents Snowden took would have never existed in the analog world. They were born digital, if you will. The digital transition has been a boon for data that is born digital and this monsoon of digital data shows no signs of abating. Email is a good example of data born digital. In most instances the information conveyed in emails starts with the email. The data only knows a digital existence. The digital nature of a medium like email and the corresponding low cost result in a much higher quantity of data creation. We share things in text messages and emails, with words and photos, that we probably wouldn't otherwise share—because of the ease of sharing, and the fact that the cost is so low.

Some of the documents leaked by Snowden describe the mass-surveillance programs of the U.S. National Security Agency (NSA) and the British Government Communications Headquarters (GCHQ). One of the surveillance programs, called PRISM, allowed the agencies access to information stored by major U.S. technology companies without individual warrants. The program also mass-intercepted data from digital communication networks. This was data born digital. It was the existence of digital devices producing digital data that made many of these documents possible in the first place. It wasn't until the advent of digital that this data could be systematically collected. Digital accelerates the creation and capture of data.

Dozens of books have been and will be written debating the virtues of Snowden's acts. This is not the place to debate whether he is a whistleblower who opened our eyes to a great injustice or a dangerous terrorist who premeditated one of the largest acts of espionage in history. Villain or hero, Snowden does play a central role in my thesis that the shift to digital has massive ramifications for data, and we've only begun to scratch the surface.

Let's now turn back to the Rubik's Cube. Although a child's toy, the Rubik's Cube is a study in order and chaos. While simple in design and function, the original Rubik's Cube (3 x 3 x 3) was advertised as having "billions" of possible states—unique ways in which the Cube could be configured. A billion is a big number, but not nearly as big as the actual number of possible states. As the story goes, the manufacturers of the Rubik's Cube, more interested in making a profit than in mathematical precision, thought that ordinary consumers wouldn't understand (or believe, perhaps) the real number, and felt "billions" sounded big enough to serve their marketing purposes.

In any event, the actual number of different possible states (rounded up) of the Rubik's Cube is 4.3×10^{19}—or, more simply, 43 quintillion. The unrounded number is 43,252,003,274,489,856,000. There are 43,252,003,274,489,855,999 wrong ways to solve a Rubik's Cube—and only one right way. As mathematician Scott Vaughen put it, if one were to physically count every possible permutation of the Rubik's Cube, moving the toy from one state to the next every second, it would "take longer than the age of the universe to reach every possible position."[3] Not bad for a cheap piece of plastic with colored stickers.

Given the literally unimaginable number of possible permutations of the Rubik's Cube, one would think that only a genius (or a super-computer) could solve the puzzle. As everyone knows, that's not even remotely true. A child with a reasonable level of determination and time can solve the puzzle. That's because the human mind doesn't need to cycle through even a fraction of the 43 quintillion possible permutations to find the one right permutation. Our brains are able to filter through the Cube's chaos to find the order—this goes here, that goes there, this might work there, aha! Done. Order from chaos.

In any one of its 43 quintillion states the Rubik's Cube is nothing but a collection of assembled blocks, seemingly without purpose. Turning it randomly, one would never in one's lifetime find the solution. It would just be chaos until the end of days. But the human brain doesn't work randomly. It sees patterns and it sees that there is a way in which the colored squares *should be arranged to make sense*. Our brain imbues the squares with volition, as if to say, "This is how they want to be." Finding that correct arrangement takes some work, but even when there are more possible permutations than seconds in the history of the universe, it's not beyond the ability of a child to do it. The one correct arrangement screams out to be found, and our brain is quite up to the task.

At the same time, as every Rubik's player also knows, it only takes two or three turns of the cube to start the game all over. That's where the 43 quintillion permutations really matter and the human brain fails. The average person can't track the shifting cubes beyond a few turns, and so, just like that, chaos from order—and another few frustrating hours for the player.

The Rubik's Cube, however, has a set number of permutations. As big as that number is, it can't increase. There are no such limitations on data, which, as I explained earlier, is infinite in volume and variety.

A UNIVERSE OF BIG DATA

Before we reach the last major development in the history of data, we need to understand fully data's cyclical nature: order from chaos, chaos from order—back and forth, from the dawn of time toward an unknown end. This property of data, more fully described in chapter one above, explains so much of what we see today. We are in the middle of a data explosion right now—and thus in the middle of tremendous chaos.

Consider some of the history of the Internet we've previously discussed. At first there are few digital documents. As the number of digital documents grew, chaos takes over. Enter Tim Berners-Lee, who invents the World Wide Web in 1989 as an information management system. The Web is born and initially the number of websites are small and manageable. But over time the number grows so large that the Web becomes a jungle of sites, and users begin to get lost in the chaos. Then in 1995, Larry Page meets Sergey Brin at Stanford. At the time Larry Page is a recent University of Michigan graduate considering Stanford and Sergey Brin is assigned to show him around. In 1997 they register Google.com—a play on the mathematical term "googol," which is 10^{100} or the number 1 followed by one hundred zeros. The name reflects their stated mission of organizing the dispersed and seemingly endless information on the web. Today we see chaos creeping back in as the data grows beyond the order in place. Search results used to contain a few hundred to a few thousand results. Today even a search for "DuBravac" returns

150,000 results and "Obama" will give you over 160 million. As data grows, we seek to create order around it, but data breaks through that order and chaos ensues until a new order can be found. This is the cycle of data—from order to chaos to a new order.

Here's another analogy: approximately 13.82 billion years ago our universe began. All matter, time, and space, even the very laws of physics, were created in that instant. As the debris expanded, filling the universe, gravity created clusters of particles, which in time formed stars, planets, and galaxies. In the swirling chaos, gravity created order. Billions of stars were born and died long before our own sun existed. But the order would give way to another round of chaos, as stars winked out and exploded. And the process will continue long after our sun disappears, its particles flung out into the universe to be the building blocks of some other star or planet. Our world, as we know it today, was born out of chaos.

Throughout the universe the cycle is repeating endlessly. Chaos breeds order, which in time begets another round of chaos. Physics teaches us that order going to chaos and chaos going to order require energy as an input. And so it is with data. Data explosions, the few that have occurred in history, have led to moments of profound chaos. Indeed, we can pinpoint just two such occasions with any sense of accuracy: the invention of the printing press and the creation of digital data. The former, as previously mentioned, was a major factor in (if not the prime mover of) the Renaissance, the Reformation, the Age of Exploration, and the Industrial Revolution. Its influence continues today, albeit in less direct form.

Here's the thing: the data that digital technology has unleashed is order of magnitude greater than that of the data released by the printing press. In fact, it's useless to compare the two because the difference is so ridiculously large. So if the printing press had a hand in *those* human revolutions, then imagine how digital data will transform the human experience. Imagine the possibilities of data unbound by physical limitations. Imagine the possibilities of data when the cost of replication gets closer and closer to zero with every moment. This is the moment in which we find ourselves. It's 1450 and the first Gutenberg Bible is hot off the press.

Over the top? Perhaps. I'll let you be the judge. But consider this. In 2013, researchers at SINTEF, a Norwegian research organization, reported that 90 percent of the world's data had been generated *over the past two years*.[4] Every second, over 205,000 new gigabytes are created—or 150 million books. That pretty much takes care of Gutenberg. But to put these numbers into their proper context—and to better understand the enormous extent of the data before us—we need some points of comparison.

According to a University of Southern California study, in 1986 the world's technological capacity to store information—both analog and digital—was 2.6 exabytes. (By the way, an exabyte is one quintillion bytes, and you now know how big a quintillion is.) To put this into context, 2.6 exabytes of data represents one CD-ROM of data per person in the world. In 1993, the number grew to 15.8 exabytes, or around four CDs per person. By 2000, it was 54.5 (12 CDs). In 2002, digital storage capacity overtook total analog capacity and by 2007, 295 exabytes of data were stored (61 CDs).[5] Between 1986 and 2007, when the study ended, the world's globally stored information experienced a compounded growth rate of 23 percent a year.

Things are just getting interesting.

The USC study looked at total storage capacity, which isn't quite the same as digital data created. But it serves our purposes for understanding the numbers that are to follow. Since 2007, the market research firm IDC has published an annual estimate of all the bytes added to the "digital universe," which it defines as "all the information created, replicated, and consumed in a single year." The size of the digital universe is measured from 2005, which for our purposes here will serve as the "first year" of the digital universe. IDC reports that thirty-two exabytes of data were created and replicated in that first year—put another way, fifteen times more data was created in 2005 than could have been stored in 1986.[6]

For the 2014 report we're in entirely different territory.[7] Forget the puny exabyte; we're on *zettabytes*—that's 10^{21} or a *sextillion* bytes. In 2013, the IDC report calculated that the world created 4.4 zettabytes of data. As the report notes by way of comparison, if the digital universe were represented by the memory in a stack of PC tablets, in 2013 the stack would have "stretched two-thirds of the way to the Moon." What's more, the digital universe appears to be *doubling in size* every two years—that's right, Gordon Moore. By 2020, the data we create and copy annually will reach 44 zettabytes— enough data to create 6.6 stacks of PC tablets to the moon.[8]

Those are staggering numbers, far beyond anything we can really comprehend without context. Fortunately, the data visualization firm DOMO calculated how much data is produced *every minute* using some familiar examples.

As of 2014, every minute…

…204 million email messages are sent;

…Google receives over 4 million search queries;

…2.46 million pieces of content are shared on Facebook;

…277,000 tweets are sent;

...216,000 photos are posted on Instagram;

...48,000 apps are downloaded from Apple's App Store;

...26,380 reviews are posted on Yelp!;

...3,472 images are pinned to Pinterest;

...and 72 hours of new videos are uploaded to YouTube.[9]

Don't worry about remembering these figures; they are already out of date—so fast is the growth of digital data that is upon on us. These figures and others just like them are getting larger each and every day. The advent of mobile technology, particularly smartphones and tablet PCs, has been the single biggest development in the growth of digital data since the Internet. Today roughly 3.6 billion people have a mobile connection and there are over 7.1 billion total mobile connections. Soon there will be more total mobile connections than there are people on the planet. Today we are adding over 400 million smartphone subscriptions every year.

According to Cisco, which publishes an annual report on mobile traffic, global mobile data traffic reached 1.5 exabytes per month at the end of 2013—up from 820 petabytes (10^{15} or a quadrillion bytes) per month at the end of 2012. Mobile data traffic in 2013 was nearly eighteen times the size of the entire Internet in 2000.[10] The world is producing more data than ever before, and we are moving it around in increasing frequency. We see the properties of data we've discussed— such as replication and movement—influencing the world before us.

Fueling this gargantuan data traffic is a relatively new player in the mobile space: video. At the end of 2013, Cisco reports, mobile video traffic represented 53 percent of all mobile traffic. Video is now possible via mobile because of the faster connection speed of 4G technology. Cisco found that mobile network connection speeds more than doubled in 2013—from 526 kilobits per second to 1,387. Even though 4G connections represent only 2.9 percent of all mobile connections, they account for 30 percent of all mobile traffic. By 2018, traffic from wireless and mobile devices will exceed traffic from wired devices.[11]

In a video announcing the 2014 IDC report, senior vice president Vernon Turner makes a familiar comparison: "Like the physical universe, the digital universe is large. By 2020, it will contain nearly as many digital bytes as there are stars in the universe: 300 sextillion. And like the physical universe, the digital universe is expanding, but much faster, doubling every two years."

To compare anything, much less something human in origin, to the physical universe is almost always hyperbole—except when it comes to digital data. In that case, the comparison is apt, for now. Within some of our lifetimes the comparison will become outdated when the digital universe exceeds the size, in terms of bytes, of the physical universe. Put another way, our children will grow up in a world where the largest thing in the universe won't be the universe; it will be the digital universe mankind set in motion.

ACCELERANTS FOR CONTINUED GROWTH OF DATA

We've touched on some of the main factors in the explosion of digital data already: Moore's Law, the deflationary effect of electronic components, and—certainly not least—the digitization of more and more things are all accelerants to continued growth in data. But this really only tells half the story and doesn't necessarily account for the exponential growth in digital data we've seen over the last fifteen years, much less the last five.

As I mentioned in the last chapter, three important elements—computational power, ubiquitous Internet access together with the analogous rise of digital communication networks, and a proliferation of digital consumer products—came together during the first decade of the new millennium. Along with the simultaneous rise in digital storage availability and a corresponding decline in the price of digital storage, these pieces together form the foundation of the explosion of digital data and will continue to work in tandem in the foreseeable future as it continues to expand. These are the four core pillars of our digital destiny (I'll introduce a fifth and final one in the next chapter). Several of them have important mathematical properties that have helped propagate the move to digital.

The first piece of this puzzle is a theory known as Metcalfe's Law. Put simply, Metcalfe's Law states that the value of a communications network is proportional to the square of the number of connected users of the system (n^2). That essentially means that if you double the number of nodes in a network—fax machines, computers, phones, and so forth—you quadruple its value. Named for the co-inventor of Ethernet, Robert Metcalfe, this theory originally applied to devices (for example, in 1993, when the theory was first postulated, to fax machines). But the Internet, and in particular social networks, have expanded its definition to include users (that is, people) as well. So, for instance, two people on Facebook can only make one connection. But five people on Facebook can make ten connections, twelve people on Facebook can make sixty-six connections, and so on.

But when you start to remember how social networks like Facebook work, then you begin to see the larger meaning of Metcalfe's

Law. One Facebook user with one friend shares a status update. Only his friend sees it, which means the data has essentially doubled. But if that friend, who has hundreds of Facebook friends, shares the update with his network, and half of them share it with their networks... well, then, you begin to see Metcalfe's Law at work. A single piece of data, via social networks and the power of Metcalfe's Law, generates thousands of replications—all in a matter of moments, and all practically free.

Of course, social networks wouldn't have much power or purpose without users, and there wouldn't be much power or purpose in being a user without a connection. The second key element to come together over the last decade is the rapid growth of ubiquitous Internet access—and, more importantly, the climb of always-on broadband connectivity within U.S. households. As late as 2000, fewer than half of all U.S. households had Internet access of any kind and only 4 percent of U.S. households had an always-on broadband connection. But beginning with 2001, adoption of home Internet access grew precipitously, and the S-shaped curve of adoption typical of innovation diffusion is evident here. By 2003, nearly 20 percent of U.S. households had broadband Internet access—a fivefold increase from just three years before. In another four years, over half of all U.S. households would have broadband connectivity for the first time. Today, broadband is the clearly preferred choice for Internet connectivity, and 96 percent of households with Internet access use a broadband connection.

Over the course of the last decade, wireless connectivity has also proliferated. With more and more public and private locations featuring free Wi-Fi, the "always-on, always-connected" dream of just a few years ago is becoming more and more a reality. Cellular connectivity has also proliferated, and today there are more cellular connections in the United States than there are people.

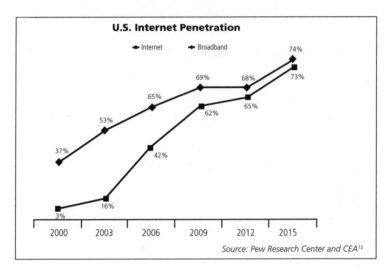

U.S. Internet Penetration

Internet access, and broadband specifically, do more than just connect these devices, making the dissemination of digital data possible. With each always-on connection, an Internet node is created and added to the network, thereby applying Metcalfe's Law to every connected device—from every smartphone and tablet in the world to a plethora of devices yet to come. The vast majority of objects that will one day connect to the Internet are not connected today. The future will see a flood of connected devices, all pouring data into the system.

The final piece of the trigger to the data big bang is the declining cost, and rising capacity, of data storage. Today, we rarely think twice about memory or storage space. Most free email systems have more than enough storage space for our needs, as do most desktops, laptops, and tablets. It's only when you're in a memory-heavy profession,

such as video and graphic design, that the worry you might not have
enough storage capacity crosses your mind.

We tend to forget that it wasn't always this easy. Consider: in
1980, a Morrow Designs hard drive with just 26MB of storage cost
roughly $5,000, or $193,000 per gigabyte. In 1985, the average cost
per gigabyte was $105,000. In 1990, it was $11,200. Five years later,
it was $1,120. Move ahead to 2015, and the average per-gigabyte
cost is under $0.05—a nickel.[13]

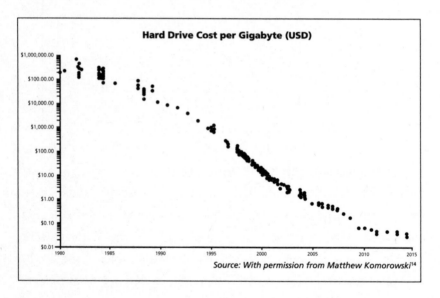

Source: With permission from Matthew Komorowski[14]

Like Moore's Law for computer processing, storage capacity
also follows a fairly stable regression curve that can be mathemati-
cally defined. As mathematician Matthew Komorowski has shown,
"over the last 30 years, space per unit cost has doubled roughly
every 14 months (increasing by an order of magnitude every 48
months)."[15] The advent and rise of cloud storage with services like
Google Drive, Dropbox, and others augment our hard drives with
storage capacities that can be accessed on any device anywhere in

the world—lowering the barrier to storing an increasing amount of data even further.

Today you don't generally consider the costs of storage—unless you're running Google or Amazon—and most consumer products carry more than enough storage capacity for the average user. Clearly, this has been a major factor in the rise of digital technology. Remember that one of the biggest obstacles for our ancestors in data dissemination was storage. As I recounted, our earliest approach involved only our memory. Later, the process of creating permanent repositories of data was expensive and laborious—paper was a luxury item and transcribing took an inordinate amount of time. Gutenberg exploded this inefficient system, but the transformation brought about by the printing press was only a matter of degree, not of kind. Books were still relatively expensive because printing presses were highly advanced machines that required skilled technicians to operate.

In other words, we have solved a problem that has bedeviled mankind since the prehistoric era. We have more than enough storage at a price per gigabyte that is cheaper than bubble gum. Yay! In fact, as Komorowski points out, we've reached the point where hard drives aren't really getting any more powerful—or at least size is no longer the primary focus. As he writes, "For a long time, one of the driving factors in consumer hard drive innovation was the need for more space, and that simply isn't a huge force anymore. Increasingly, we're focusing our attentions on speed, reliability, and accessibility. There are several external forces that have shifted the focus away from the mantra of "more space cheaper," and it's hard to imagine the industry ever pivoting back.[16]

For starters, we simply don't need as much storage on our personal devices as we used to. Companies like Netflix exploded onto the scene with video streaming, and the rise of cloud storage has only accelerated the trend. We don't keep things locked in our hard drives; instead, we let services like Dropbox store them for us, just as a bank stores most of our money. As we'll see in a moment, this hasn't been entirely for the better.

Nevertheless, throughout the first decade of the new millennium, the cheap cost of unlimited storage catapulted our ability to disseminate data at exponential levels and speed. Matched with the compounding nature of Metcalfe's Law and the "always-on, always-connected" reality of today's devices, solving mankind's storage dilemma was a key accelerant in the growth of digital data.

In the data explosion we are currently experiencing, diverse and disparate forces come together. In many ways, that is a common story of inflection points. Massive change takes place when a myriad of forces come together. A dry forest. A lightning storm. A spark in just the right place at just the right time. All of these factors were working towards this day. Data has been struck by digital and the landscape is forever changed.

THE DIGITAL DISRUPTION

In almost all cases, digital magnifies the properties of data that we discussed in the opening chapter above.

As you'll recall from that earlier discussion, one of the characteristics of data is its infinite divisibility. But in an analog world there are often constraints to the infinite divisibility of data. There are demands by a mass market to find something that has mass appeal; there are shelf space considerations in a store; the clock of the analog TV broadcast was limited to 24 hours a day, 1440 minutes, 86,400 seconds. These were real limits that had to be addressed daily by everyone working in the analog world.

These limits are why we often see bundling employed. The hope in the analog world was that by pulling different pieces of information together you could pull together enough to create broad mass appeal even if everyone making up the mass didn't find every part of the bundle appealing. Take for example the evening news, which bundled together weather, sports, and local and national news into a standard thirty-minute television broadcast spot. On the surface these are all pretty different topics that appeal to very different audiences. I recall as a kid having to suffer through local stories that didn't interest me at the time in order to get to the evening's sports highlights. In many ways, cable television tried to overcome this scarcity-driven-bundling by offering more channels, which could then cater to well-defined niche needs but in aggregate attract a mass market. In the world of cable you could offer a sports channel, a channel dedicated to home remodeling, a channel dedicated to cooking. But cable didn't fundamentally overcome the scarcity problem.

In digital, many of the limits imposed on data by bundling are not warranted. I don't need to sift through other news of the day to see all of the sports highlights I want. I don't have to watch a morning talk show only because I like the cooking segment.

Look at what happened when music met digital. Before the advent of digital, music was curated by record labels who decided what would be pushed to the masses and what was "not good enough." In the analog world, shelf space was a scarcity, so record labels had to use it judiciously. There was an opportunity cost—if the label picked something that didn't sell as well as something else would have, they missed the opportunity of the higher sales. Bundling also helped cope with the uncertainty about what would be popular by putting several songs together in the hopes that one of those songs would be popular enough to induce purchase.

When you aren't sure which songs will be popular, you put four-teen or fifteen songs together (on an album) in hopes that some buyers will like one song while others might like a different song. Collectively all interests are served, and a mass market is created. We see this same thing in other industries. Physical newspapers bundle several diverse stories together to create a product that appeals to a wide audience in order to overcome the problem of shelf space scar-city.

With digital data, you don't face the same cost constraints that you did with analog information. There's no shelf space scarcity, so the music industry can break up the individual tracks and sell them separately. The newspaper can offer access to just a single story, without customers having to buy every story. All of a sudden, digital breaks down limitations and boundaries that were put in place because of restrictions related to analog: articles of a certain length, books of a certain size, movies of a certain genre. As a result, the ubiquity of digital opens up a huge swath of offerings that weren't available because of these limitations, and we as consumers benefit tremendously.

Editorial columns—previously defined by the analog size of the broadsheet newspaper page size and layout used—are freed of word count limitations. In a world made up of digital, filled with blogs and countless online content, size doesn't matter. Books of a certain prescribed length are no longer the only option. In January 2011—as I was just beginning to put the pieces together for this book—Amazon launched Kindle Singles, which opened up an entire marketplace for books below the previously prescribed lengths.

On YouTube and across the Internet today, you can find millions of videos of people doing all sorts of activities—from fixing things to playing video games. (Amazingly, people can amass followers who will watch them play recordings of themselves playing video games.) This would never happen in a world with shelf space scarcity.

When we take down the limitations, the number of offerings explodes. This is especially true in the creative industries such as music, movies, writing, and the arts. Digital platforms for distribution are allowing the provider to connect directly with the consumer.

We see the same thing happening in services. Where narrowly defined services were not viable in an analog world because they could never attract a large enough market, digital services can scale across a much wider geographic base. The great digitization of data is impacting not just information but also services. You can create a shirt from a million different possible designs thanks to services like CafePress. I've made countless shirts on CafePress—to give as presents, or for myself. On very rare occasion some of the designs were appealing to others who then bought them also. But none of them was available before digital because they would not have been given the shelf space. The digitization of markets for things like t-shirts and mugs makes markets of single units viable markets because of the infiniteness of digital shelf space.

Thanks to digital, we end up with a Rubik's Cube of options.

CHAPTER 4

The Sensorization
of Objects

"A good decision is based on knowledge and not on numbers."

—Plato

I n 2013 I made my first trip to Ethiopia—and my first trip back to the African continent since 1999. Knowing a bit about the country's economy and surroundings, I expected the grim poverty that I encountered. There are encouraging trends in Ethiopia, such as better water, more children in schools, and rising per capita income. But many of these improvements are relative. For instance, the per capita income in Ethiopia is a mere $380. Ten years ago, it was $130. Similarly, the country's 29 percent poverty level is far better than the 38 percent from a decade earlier.[1] Nevertheless, Ethiopia remains an impoverished nation dependent on international assistance.

So imagine my surprise when I entered a restroom in a small town outside of Addis, the capital, and found urinals with sensors—the kind that know when you're, um, finished. Believe me, in the U.S. and most of the places I travel I don't normally notice things like urinals. But in Ethiopia, where indoor plumbing is considered a luxury and a lot of folks live on less than a dollar a day, self-flushing urinals catch your attention.

It's not as if modern technology is absent from Ethiopia; it's just at nowhere near the level of penetration it's at in wealthier nations. For example, there are 20 million active cell phones in Ethiopia, according to the *CIA World Fact Book*.[2] But in a country of 90 million people, that's a mere 22 percent penetration—far below the near 100 percent levels we see in the West.

And of course a single restroom with a few sensorized urinals doesn't mean much in the grand scheme of things. Ethiopia needs many more basic technological upgrades just to reach what we would consider an acceptable standard of living. To find something as relatively advanced as a sensorized machine in Ethiopia doesn't say much about the country; but it says a lot about the machine.

In particular, it says a lot about sensors. Do me a favor right now. Take out your smartphone. Keep it nearby as you read the next few paragraphs. There's actually a pretty good chance you are reading this book on your smartphone. Depending on what kind of phone you own, there are between five and nine sensors inside it, and the list is growing by the day. It's worth spending a moment to run through the most common ones because each of these sensors is digitizing data as we speak:

Proximity sensor: Usually located near the hearing speaker, this one recognizes when the device is brought near the face during a call and deactivates the display and touchscreen. It helps you save battery life and avoid inadvertent touches while talking on the phone.

Ambient light sensor: This one simply recognizes the amount of light in the phone's immediate vicinity and adjusts the display brightness to save power and make for a more comfortable reading experience.

Accelerometer: Now we're getting to the cool stuff. Today's 3-axis accelerometer, embedded inside the phone, senses the orientation of the phone and changes the screen accordingly. Not only does this help with reading and viewing images, but games also make use of the accelerometer to mimic player moves.

Gyroscopic sensor: A sensor found on newer smartphone models, the gyroscopic sensor works like the accelerometer and measures rotation. Gyroscopes increase the degrees of motion that can be measured and tracked.

Magnetometer: If you have a newer iPhone then you likely also have the Compass app, which uses the device's magnetometer to detect the strength and/or direction of the magnetic field in its vicinity. If you haven't used the Compass in a while, open it up now. It'll likely ask you to "calibrate" by rotating the device in a circular motion. That's the magnetometer sensing the earth's magnetic field (or a stronger one nearby).

Ambient sound sensor: If you're in a crowded room, you might feel the need to step outside to talk on the phone. But that's the problem the ambient sound sensor, usually located on the back of the smartphone, is meant to solve. It takes in ambient sound—basically, everything other than your voice—and cancels it out, so that the only thing the person on the other end of the phone hears is your voice.

Barometer: Available on some Samsung smartphones and the iPhone 6 released last year, the barometer does what you'd expect: measures atmospheric pressure at your precise location. While this does lead to more accurate weather forecasts, the true value in the barometer is in health-based apps. A running or walking app, for instance, would be able to give a more precise count of your calorie burn if it also knew the pressure (or altitude) at which you were exercising.

Temperature/humidity sensor: Found on some Samsung models, this sensor works with apps to provide better health readings in an exercise environment—and, yes, better weather forecasts too.

M7 motion coprocessor: A feature on the iPhone 5s, the M7 combines the readings of the accelerometer, gyroscope, and compass to give your phone even more awareness of its (and thus of your) surroundings. The M7 knows whether you're walking, running, or driving in a car and feeds this information to the phone's processors, which adapt its activity to what you're doing.

Today you probably take most of those sensors for granted. How much do you think all that high tech costs? The costs of all the sensors add up to under $5.00. Some obviously are more expensive than others, but the cheapest can be bought for as little as $0.07.

Now you know why a poor country like Ethiopia can afford a luxury like a self-flushing toilet. Sensors, the last development in our march into the future, are dirt cheap. And yet they are revolutionizing the human experience.

FROM AIRBAGS TO OCULUS

In 1997, futurist Paul Saffo wrote that the 1980s were the "microprocessing" decade, because at that time our personal computers were judged on their power to process—a solitary exercise that depended on what data a human being fed into the machine. In the 1990s, he wrote, the focus shifted from processing to access. Cheap lasers slipped into every corner of our lives, from CDs to the fiber-optic cables that delivered broadband. Because of the cheap storage supplied by CDs and the speed with which fiber-optic phone lines delivered broadband, our attention turned from what computers could process to what computers connected us to, in other words, access. Computers were now networked devices.

Looking ahead, Saffo wondered what next tech development would define the coming decade. He chose sensors:

> Hints are lurking in many areas, but one of the most intriguing indicators appeared in Los Angeles in the last two years. What is the most popular item to steal out of automobiles in Los Angeles today? Air bags—because they contain an expensive and not-entirely-reliable accelerometer trigger. The consequence has been a booming market for replacement airbags, which thieves are happy to fulfill.
>
> Air bags are about to become too cheap to steal, however, because, using MEMS (MicroElectroMechanical systems) technology, one can build an accelerometer on a single chip for a couple of dollars, creating a device that is not only cheaper than today's sensors, but also smarter and more reliable. Today's systems dumbly explode whenever they sense an abrupt acceleration, whether or not a passenger is present. Future systems will incorporate sensors capable of identifying not only the presence of a passenger, but their weight and size as well, and adjusting the force of inflation accordingly.[3]

I love that phrase: "too cheap to steal." It perfectly captures one of the central ideas of this book: namely, that our digital destiny has been driven not by monstrous, delicate, billion-dollar hardware, but by cheap electronics and even cheaper plastic. It also highlights another important point. Often our next tech wave comes from places we would never have considered at the time. Who, other than a futurist, would look at airbags and say, "That's our future!"?

In any case, Saffo was right—airbags became too cheap to steal. So let's give him another word or two:

Such new devices—cheap, ubiquitous, high-performance sensors—are going to shape the coming decade. In the 1980s, we created our processor-based computer "intelligences." In the 1990s, we networked those intelligences together with laser-enabled bandwidth. Now in the next decade we are going to add sensory organs to our devices and networks. The last two decades have served up more than their share of digital surprises, but even those surprises will pale beside what lies ahead. Processing plus access plus sensors will set the stage for the next wave—interaction. By "interaction" I don't mean just Internet-variety interaction among people—I mean the interaction of electronic devices with the physical world on our behalf.

The Shift from Processing to Access

Processing	Access	Interaction
Personal Computer	World Wide Web/ Internet	Smartifacts
		Sensors
	Laser	
Microprocessor		

| 1980 | 1990 | 2000 | 2010 |

Source: Paul Saffo, with permission

As you can see from the diagram, Saffo's word for these sensorized devices at the time was "Smartifacts." Adding sensors not just to electronic devices but to everyday objects is a big part of our digital destiny. When you hear talk about the "Internet of Things," that's what is being referred to: common household objects are "sensorized" and connected to the Internet, over which they share relevant data with relevant systems. But let's not get too far ahead of ourselves just yet.

Sensors round out our history of data. Sensorization is the obvious ending point for the story to date because we are just at the start of the sensor revolution—or sensor age, if you prefer. Airbags were just the beginning. Sensors make driverless cars possible. Sensors help turn our mobile phones into "smartphones." Sensors have created the latest wave of "wearable tech." Sensors are the reason Apple has hired a slew of tech folks from the health field. As a 2013 story reported, "Based on new hires, it seems that Apple's interest in sensors focuses on the ability to measure glucose and other body level information. With this data, the product could inform users of vital information in a non-invasive way. These sensors could also pick up more data to give a user a snapshot of their health, which would be ideal for fitness applications."[4]

Or, as Apple CEO Tim Cook said at a conference in 2013, "The sensor field is going to explode." I'd take that prediction to the bank.

Seeing where we've come on our long road toward releasing the power of digital data, sensors are the next obvious evolution along our continuum of technological advancement. If you recall what we saw in chapter one about how the history of data has been one long attempt to recreate the brain's data processing power, then sensors make complete sense. Our brains take in the world around us at an astonishing rate. Our ability to process innumerable bits of data at a moment's glance hasn't been equaled by anything manmade. Of course, the technology industry is working on it, but today the only thing that comes close is the occasional super computer, like IBM's supercomputer "Blue Gene." Even then, Blue Gene can only simulate 4.5 percent of the brain's neurons and the connections among them, called synapses—that's about one billion neurons and 10 trillion synapses.[5]

So, it's going to take a while. But in the meantime, sensors, applied to everyday objects, are the next best thing. You'll recall from the last chapter that I referred to ubiquitous computing, connectivity, proliferation of digital devices, and cheap digital storage as the four pillars of our digital destiny and at that time I promised you the final pillar. Well, now you have it: sensors. The first four drove us partly into our digital destiny, but sensors cause the real acceleration of digital data. The first four pillars connect sensors together, cache their data, and help us make sense of it. If the goal of computing can be simplified as an attempt to mimic elements of the human mind, then sensors play a vital role. Sensors take in the world around them and feed that information to a processor, which helps turn that raw data into intelligence. The beauty of sensors is that they do this autonomously. In other words, they don't require human effort to input data—a laborious and time-consuming process. Thus, another barrier keeping data confined is washed away. We'll look at the ramifications in the next section below.

It would be a mistake to try to draw a direct line from Saffo's airbags to today's smartphones. The airbags that Saffo mentioned were merely an example of the trend he predicted. He saw "cheap, ubiquitous, high-performance sensors" in one industry and envisioned that they would eventually spill over to many other industries. The phenomenon of sensors in consumer electronics devices has a long pedigree, which we don't need to detail in full. From the computer mouse to the microphone, sensors have long been a staple of consumer tech, particularly computers.

It was in games, however, that sensor technology saw real leaps. Sensors in gaming go back a long way. In fact, the first light game gun, called Ray-O-Lite, appeared in 1936. In that game, players shot a beam of light at moving sensors (ducks, as a matter of fact) on the "screen."[6] But this version, like most others that came after it, was an "all-in-one" system—the screen and the gun were the game. That changed in 1985. In chapter two, we saw that in 1985, coincidentally at the International CES, Nintendo released the NES, whose first shipments included the Zapper, or light gun. I have fond memories of blasting those poor ducks in the classic game, "Duck Hunt," but at the time I had no idea how it worked. It seemed to be an amazing technology at the time, considering the screen I was shooting at had no connection to the game console itself: unlike those early arcade shooters.

Here's how it worked: The TV screen didn't know when you pegged a duck. The Zapper knew. When you depressed the trigger on the gun, it caused the screen to go black, except for the areas around the ducks, which turned white. The sensors in the Zapper detected the light variation. A dark reading meant you missed. A light reading meant you hit a duck. It was that simple. But at the time it was quite a revolutionary product and proved a huge hit for Nintendo, which went on to dominate the game console market for nearly a decade.

Nintendo would go on to introduce other early sensor-populated devices. In 1989, Nintendo released a controller accessory to the NES called the "Power Glove." Relying on ultrasonic receivers and speakers, the Glove essentially used motion sensors to detect movement. Sensors were also embedded in the finger sockets, which sent signals directly to the system through a cable.

The idea of the Power Glove—using motion to control a game—would eventually pay off for Nintendo, seventeen years later. In

2006, Nintendo released the Wii console, whose primary feature was a new kind of remote control—a wireless handheld device which detects movement in three dimensions through built-in accelerometers. With the Wiimote, as it is called, Nintendo hit another home run commercially while extending the trend toward sensorization. The Wiimote also introduced consumers to the power of sensors by moving them from behind the scenes in places such as airbags to consumer-facing applications.

It should be no surprise that sensors were an early mainstay of gaming. Gaming has often been the playground of the consumer tech world, where niche products "cut their teeth," so to speak, before maturing into more practical devices. From the Zapper to Microsoft Kinect, using sensors to enrich the gaming experience has been a key trend driving us towards the digitization of everything. Today we see this march forward continuing with virtual reality (VR). The first games utilizing virtual reality, otherwise known as immersive media, appeared in the early 1990s. Early versions were too costly, not to mention too unwieldy, for the home market. Players needed to wear a heavy, awkward headset to play. The graphics were also poor in quality and the games a bit too simple for a generation that already expected more. Ever the innovator, Nintendo released a home version of a virtual reality system, called the Virtual Boy, in 1996. Although affordable at $180, the system suffered from the same problems that had plagued earlier versions of virtual reality games: the huge headset, bad graphics, and games that didn't appeal to broad markets.

After that, VR mainly disappeared from the home gaming market. Fearful of failure, the industry stuck with what it did best: produce engaging games with great graphics. But it was only a matter of time before virtual reality would continue its march forward. In

March 2014, Facebook announced that it had bought the virtual-reality startup Oculus VR for $2 billion. Oculus's one product was the Rift—a VR headset for 3D gaming. At the 2014 International CES, Oculus showcased a consumer version of Rift, codenamed "Crystal Cove," and early reviews were quite laudatory. Indeed, the headset won the 2014 Best of CES award from Engadget.[7] The sensors in the headset track both the head and body and resolve another problem that troubled earlier VR systems, namely motion sickness. As with so many of the technologies I get to see firsthand at CES, my experience with the Oculus Rift was incredible.

Although Oculus hasn't set a release date for the consumer Rift, NASA is already using a version to simulate "walking" on Mars.[8] Other industries are also excited about the possibilities of Rift, as Engadget notes: "With the latest Rift, Oculus has created a device that may usher in an era of truly immersive gaming and entertainment, and even create new opportunities for businesses to use virtual reality in everything from manufacturing to medical environments."[9] As Mark Zuckerberg wrote on his Facebook wall when the acquisition was announced, "[Facebook's] mission is to make the world more open and connected…at this point we feel we're in a position where we can start focusing on what platforms will come next to enable even more useful, entertaining and personal experiences.…This is really a new communication platform. By feeling truly present, you can share unbounded spaces and experiences with the people in your life. Imagine sharing not just moments with your friends online, but entire experiences and adventures." Whether Rift is a success or failure, the trend is clearly headed in one direction. With earlier experimentation on sensors in games, early kinks were resolved and possibilities realized. These sensorized devices have all pushed sensor development far beyond where it might otherwise be. They also raised consumer expectations for other segments of the consumer tech industry and other industries that manufacture products for consumers. When a consumer plays with a Wii, he asks why his other devices and objects can't have the same functionality and digitize the physical space around him.

Next up we'll examine the link between sensors and data and show how the two are aligned. It is this relationship that will define a great part of our digital destiny.

THE RISE OF SENSORS

At the 2014 International CES, Jim Farley, Ford's Global VP/ Marketing and Sales, raised some eyebrows with a comment during a panel discussion. "We know everyone who breaks the law, we know when you're doing it. We have GPS in your car, so we know what you're doing. By the way, we don't supply that data to anyone."[10] Well, *that's* good to know.

Farley's admission, surprising as it may be to some, is entirely obvious. *Of course* Ford and other car manufacturers are sensorizing everything in our automobiles. Why wouldn't they? If the technology is there, and it is, it makes sense for them to want to compile as much information as they can about an individual car's performance. Sensors digitize information about the vehicle, making valuable data now easily accessible and digestible to the car's computer. This data helps Ford mechanics diagnose problems with your car more easily, and it helps Ford make better cars for tomorrow. As sensor prices decline, they are also becoming more heavily utilized in consumer-facing applications, meaning that sensors will eventually influence our entire driving experience.

It is worth noting that cars have been using sensors for decades. But until recently, these sensors operated locally. They operated only within a closed loop. Now devices that connect into the ODB-II port of your vehicle can watch this data and provide new value and ulti-mately new services. It is the connectivity of the sensorized data that was largely missing until now.

In any event, now we get to the real significance of the sensoriza-tion of everyday things. It's not so much that these sensors are responding to stimuli; it's that these sensors are feeding data to be processed. In other words, they're capturing data. And lots of it.

According to the "digital universe" report from IDC (Interna-tional Data Corporation), "data just from embedded systems—the sensors and systems that monitor the physical universe—already accounts for 2% of the digital universe. By 2020 that will rise to 10%."[11]

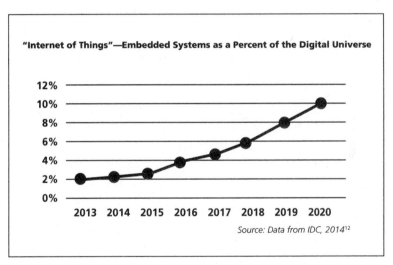

"Internet of Things"—Embedded Systems as a Percent of the Digital Universe

Source: Data from IDC, 2014[12]

Indeed, the IDC report makes special mention of the rise of the "Internet of Things." It identifies three "growth spurts" in the digital universe in modern times. The first was when digital camera technology replaced film; the second happened when analog tele-phony went digital; and the third, when television went digital. In this each of these, you can see the four core pillars of digital data that I've outlined in this book: greater computer power, growth in digital storage, connectivity, and the digitization of consumer products. All three growth spurts greatly expanded the digital universe and played key roles in the transition to digital that we discussed earlier.

As the IDC report notes,

> Now comes a fourth growth spurt—the migration of analog functions monitoring and managing the physical world to digital functions involving communications and software telemetry.
>
> Call it the advent of the Internet of Things (IoT). Fed by sensors soon to number in the trillions, working with intelligent systems in the billions, and involving millions of applications, the Internet of Things will drive new consumer and business behavior that will demand increasingly intelligent industry solutions, which, in turn, will drive trillions of dollars in opportunity for IT vendors and even more for the companies that take advantage of the IoT.[13]

To bolster this claim, the IDC report includes some eye-popping statistics. For instance, IDC estimates that the total number of "connectable things" in the world is around 200 billion. Of those, around 20 billion are actually wired and talking to the Internet now. They are able to do so through a network of around 50 billion sensors that track, monitor, and feed data to those connected things. By 2020, however, the number of "connected things" will grow by 50 percent, to 30 billion. And the network of sensors? We're talking about a trillion or so.

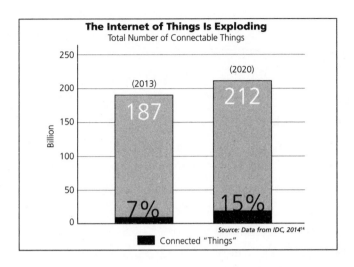

The Internet of Things Is Exploding
Total Number of Connectable Things

(2013) 187

(2020) 212

7%

15%

Billion

Source: Data from IDC, 2014[14]

Connected "Things"

The explosion of sensors is made possible because of the foundational infrastructure of the four core pillars of digital data and the declining costs (and improving quality) of sensors. Sensor prices have followed a similar trajectory to that of computer hard drives in terms of power and cost. In 2007, for example, when accelerometers were just starting to show up in video game controllers and smartphones, measuring a single axis of motion cost around $7. Accelerometers capture two planes of motion; therefore, to add the simple function of being able to change the screen orientation on a device such as a smartphone from landscape to portrait at 2007 sensor prices would have run the price up about $14. Today an accelerometer that can capture the same axis of motion costs less than $0.50—or under a buck—to add the functionality to a device. The steep price drop is a function of both strong competition in the smartphone arena and the growing number of applications utilizing the technology. The same phenomenon will likely play out in other sensor categories, such as pressure sensors and moisture sensors, as their implementation increases.

In a 2010 piece on *EETimes*, Walden Rhines, Mentor Graphics CEO, pointed out a pattern:

> One of the more amazing aspects of the increasing pervasion of semiconductors into new applications is the significant growth in revenue of existing applications as the cost per unit decreases. Consider the digital camera. Most of the semiconductor content of a digital camera consists of non-volatile FLASH memory and the image sensor. In the early 1990s, solid state image sensors sold for $20–25. Image sensors were a negligible portion of the semiconductor total available market (TAM) until the current decade. During the 1990s the price per sensor fell dramatically from the $20–25 range to about $5. At this price point, unit volume soared, making image sensors more than 3 percent of the semiconductor TAM in the last few years.... The result was a substantial net growth in the market for digital cameras and the semiconductors required to make them.[15]

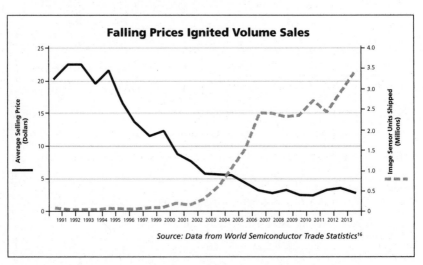

Falling Prices Ignited Volume Sales

Source: Data from World Semiconductor Trade Statistics[16]

If you look at the graph, you'll notice that even though sensor prices had been falling quite steeply since the early 1990s, it wasn't until the early 2000s that the exponential growth in the units shipped really took off. Early on, high-end phones added cameras, pushing up volume and driving down price, making it possible to embed image sensors in cheaper phones. As phone usage exploded, image sensors rode the coattails. It was at this time that prices were low enough to warrant embedding the technology into a wider diversity of devices and including multiple cameras on a single smartphone.

This is a very common trend within technology—but especially with sensors. In the early life of a given technology it is often relatively expensive, and so the given technology is used sparingly. For example, prior to 1984 computer processing power was a scarce resource and so it was used sparingly. Prior to 1984 computer manufacturers would have never "wasted" computer power on things like rendering a graphical user interface (GUI)—the icons on a screen that we are so accustomed to using today. Rather, manufacturers saved the scarce processing power to perform other more highly valued computations. A GUI just depicts graphically what you could get through by typing the code prompts we relied on prior to 1984, so not only does it cost processing power, it is also redundant. But around 1984 this all began to change. Processing power moved from being a scarcity to a surplus, so we began to "waste" it by creating GUIs. In 1984 Apple's original Macintosh became the first commercially successful computer with a GUI, and we haven't looked back since.

Digital storage is another technology resource that was once a scarcity. Now we back up multiple copies of the same file and don't give it a second thought—just to save tens of thousands of photos on our computers that we will probably never look at again. Once a technological resource gets cheap enough, we tend to "waste" that resource on noncore applications. We tend to create redundancies, such as including forward-facing and backward-facing cameras on the same smartphone. When sensors were expensive, we used them sparingly. But as prices dropped, we started to include them in a growing number of devices across a wide assortment of applications.

A SENSORIZED WORLD

In the pages that follow, we're going to be returning to sensors again and again, so at this point there's no need to go through the numerous sensor applications that are being defined. Let us instead return to where we started this chapter, in Ethiopia. Finding a sensor urinal in a distant corner of Ethiopia doesn't necessarily promise a high-tech future for Ethiopia. The cost of sensors is now cents on the dollars. And that's the truly remarkable thing. Most of our discussion up until this chapter has dealt with highly advanced electronic devices and systems. While normal in the West, these systems would still seem alien in a country like Ethiopia.

But sensor technology is accelerating toward a point at which they won't be high tech at all. Is a sensor urinal high-tech? Perhaps it still is in some places, but in time it won't be—just as, back home, the sensorization of everyday objects will become so commonplace that we won't see these "connected things" as high tech. They'll just be things. Take for example the proliferation of motion sensors in use today. We don't give it a second thought when we approach the door of a retail store and it opens as we approach it. I'm sure the first door to integrate motion sensors felt a little like magic. And that is what technology does. It takes the magical and makes it mundane. This, in a single example, is the long march of technological progress: making the remarkable ordinary—even invisible. The world isn't getting more high tech, the tech is meeting us down on the ground. It's becoming commonplace, as ordinary as the hammer in your toolbox—in the great democratization of technology.

We tend to think that our future will be one big electronic complexity, but the reality is going to be much less complicated. It's still going to be "advanced"—in that the things we will be able to do would surprise our younger selves—but the technology won't be obtrusive or arduous. That's because these dirt-cheap sensors will be enabling much of the complex work for us. They'll assume the heavy data lifting, up till now performed by human beings, while we'll go on with living our lives. Even in a country with comparatively limited opportunity like Ethiopia, our digital destiny is unavoidable.

CHAPTER 5

Second-Order Effects of Digital Data

"We don't need any more Fart apps."

—**Apple's App Store Review Guidelines (September 2010)**

Thus far we've primarily looked at a number of the first-order effects of digitizing data. The first-order effects are the initial effects of digitizing information—things such as greater access to information and greater abundance of available data. As I discussed earlier, Moore's Law brings us lower prices for digital devices and opens up greater access to technologies. As more markets and services digitize, greater access to technology gives us greater access to these markets and broader access to digital everything. As we saw in the last chapter, the sensorization of consumer technologies will open up even greater customization and personally relevant experiences for us in the decades to come.

But these effects also become causes for other effects continuing on in a long cascade of effects and causes. The second-order effects of digitization are the outcomes caused by the first-order effects: greater abundance and great access. These unintended consequences create challenges that will need to be addressed as we increasingly move into an all-digital world. Digital brings us lower prices, which bring us more choice and greater selection. This is true for all things—not just for the things we want, but also for the things we might not want more of—from the preponderance of digital cameras tracking our every step to advertising tailored to our every move. Digitization equally applies to things we may have but do not want more of. Digital makes communication easier than ever, for example. It means we can email anyone we want in the blink of an eye. But it also means that it is relatively easy for anyone who has our contact information to email us in the blink of an eye as well.

The marginal cost of a digital copy is close to zero. In the case of email, for example, this means the additional cost of sending another email is practically zero. Because the cost approaches zero, the market becomes oversupplied. The end result is not only quick communication with others, but also widely distributed low-value email advertising: SPAM. Good content can be lost in the tidal wave of bad. It takes only a few short steps to upload a video online. With the great proliferation of smartphones that are continuously connected and have recording capabilities, it has never been easier. This has created an oversupply of videos across a wide selection of genres. Perhaps not all broad genres, but take things down a few levels and it is easy to see the potential for excess. How many videos do we need of a kitten chasing a laser pen? Or of some nonspecific firework show?

In almost all cases, when a market digitizes, it changes from a market defined by scarcity to one defined by abundance. Purely in reaction to oversupply—the flood of low-value material overwhelming the good—we attempt to curate our numerous experiences navigating digital markets. Look at video, for example, where we've created online video "channels" in an attempt to curate content and form order from the massive amount of choice that is provided through the digitization of video. We are in a constant struggle to curate value wherever digital markets create abundance.

While I'm personally grateful for the more than 125,000 videos that show up when I search for "MG Midget," the superfluity also hampers everyone's overall experience—including mine. While digital makes "markets of one" viable, it also means I may have to cull through dozens of these markets to find the one that is right for me. Search and other tools were designed to help cull through these marketplaces and retrieve what you wanted without having to do all of the heavy lifting. But while it helps on the margin, advanced search functions only get you so far.

About a year before Steve Jobs died, Apple published revised App Store Review Guidelines. At the time, Apple's App Store had about 250,000 apps, and the revised guidelines suggested that the company was going to be more judicious about what apps they allowed in the future. In other words, they weren't going to digitize something just because digital allowed it. Here are some of Apple's guidelines:

- We have over 250,000 apps in the App Store. We don't need any more Fart apps.
- If your app doesn't do something useful or provide some form of lasting entertainment, it may not be accepted.

- If your App looks like it was cobbled together in a few days, or you're trying to get your first practice App into the store to impress your friends, please brace yourself for rejection. We have lots of serious developers who don't want their quality Apps to be surrounded by amateur hour.

In a digital age, the risk of drowning in bits is real. Apple was making an effort to throw us a lifesaver.

DATA'S CHAOS-ORDER CYCLE

In the summer of 2012, a rather significant discovery was announced from the scientists at the European Organization for Nuclear Research, better known as CERN. Using a machine called the Large Hadron Collider, the world's largest particle collider, scientists at CERN smashed two proton beams together to reveal a mysterious particle known as the Higgs boson, which was first theorized to exist nearly forty years ago. Among other things, the Higgs boson could help explain how the universe expanded immediately following the Big Bang. With matter spreading in all directions, its very movement creating time and space, the early days of the universe were anything but ordered and tidy. Even now, 13 billion years later, the universe has yet to "settle" and likely will only settle when it stops expanding and the whole big thing burns out. Until then, the universe is a mixture of order and chaos—order being created in one sphere just as it unravels into chaos in another. That's much like the current state of digital data, only there is far more chaos at the moment than order.

Digital was the Higgs boson of data. It has unleashed on the world a level of data akin to the Big Bang. With this data, of course, has come great chaos. Each data revolution unleashes a maelstrom of information, but in time humans find ways to put in place a sense of order. Then, when more data is unleashed, the process repeats itself.

The volatility in this cycle between chaos and order impacts the decisions that individuals make and will be an ongoing theme with respect to digitized information. Despite Americans' overwhelmingly positive attitude toward the Internet (and by extension digital data), there remains an undercurrent of deep-seated concern. For instance, 86 percent of Internet users, according to Pew, have taken steps to "remove or mask their digital footprints," such as deleting cookies or encrypting email. It's no surprise why. More than a fifth of all Internet users have had an email or social networking account compromised. Smaller percentages have had more negative experiences. In the words of the Pew study:

- 13% of internet users have experienced trouble in a relationship between them and a family member or a friend because of something the user posted online.
- 12% of internet users have been stalked or harassed online.
- 11% of internet users have had important personal information stolen such as their Social Security Number, credit card, or bank account information.
- 6% of internet users have been the victim of an online scam and lost money.
- 6% of internet users have had their reputation damaged because of something that happened online.
- 4% of internet users have been led into physical danger because of something that happened online.
- 1% of internet users have lost a job opportunity or educational opportunity because of something they posted online or someone posted about them.[1]

So even though we're talking about very small percentages, eight in ten Americans nevertheless feel as though the Internet isn't completely safe. But is that a bad thing? It depends. The Internet isn't, in fact, all that safe. You can get scammed, hacked, or infected with a virus quite easily if you don't know what you're doing. The dangers of the Internet have only grown as it has become a larger part of our lives. As the danger has grown, so have our paranoia, suspicion, and skepticism, but so too has our reasonable wariness.

The nuance here is that every place has risks. I recall driving in Nigeria and noticing that the gas tanks on almost all of the trucks I saw on the road had often-crudely-welded metal over the gas knob with a padlock, presumably to deter would-be thieves from attempting to syphon off the gas. The frequency with which I observed this method of securing a tractor-trailer's gas tanks suggests that gas theft must have been problematic enough or posed a significant enough risk to warrant the widespread application of the solution. But obviously the risk wasn't so severe that it stopped truckers from trying to carry out their daily activities.

Risk management for digitized data works in much the same way. Sure there are risks. There are risks today to driving in your car or eating certain foods. These are risks we are aware of and assume. But we also take precautions to avoid any large negative outcomes related to these risks. We wear seat belts and wash fruit before we eat it. We weigh the cost of precautions against the potentially large negative outcomes. The risks and rewards of data, and especially digital data, are embedded in the cycle between order and chaos. Chaos—driven by abundance—creates risks and uncertainty that we try to manage by creating order.

From the Internet's early days, we all remember those email scams that asked us to claim untold millions of dollars from some foreign nation—we just had to send our bank account information. Few people fall for these anymore. Regardless, those early impressions left a mark on our young Internet brains, which is one of many reasons why so many of us take pains to remove our "digital footprints," as futile as that can be.

Suspicion isn't always a bad thing. With 50 percent—up from 33 percent in 2009—of users saying they are worried about the amount of personal information about them that is online, we can imagine that these users display more caution in their online interactions. As the Internet has become more central to our lives, people have generally responded with a higher level of guardedness.

Using the Internet as a proxy for digitized data in general, we can start to see how we might manage the large number of diverse aspects of our lives that will be impacted as they collide with digitization. It takes energy to create order from chaos in the physical world, and the same holds in the digital world. And we humans put a lot of energy into putting data into order…right up until the next chaos-inducing event. Got your database of MP3s in order? Great, here's Spotify. This cycle continues in every arena, and it is pounding the aspects of our lives currently being consumed by digital.

It is simply impossible for the magnitude of data that has been created in the last fifteen years to have anything less than an unsettling effect on a world that was entirely analog for so long. In certain cases, we have grown accustomed to the chaos—"this is the world we live in" being a common refrain to explain a phenomenon that simply didn't exist only a few years ago. In some places, the chaos has given way to order, as we develop new ways of dealing with newly digitized data. But we're still a far way off from anything resembling the predictable order that preceded the digital age, and the consequences of this chaos are everywhere around us. Let's continue to consider some of the most significant examples.

DATA BREACHES

Data breaches have long been a part of data's history. In the 1920s the German military began using ENIGMA machines to encode radio messages. The ENIGMA machine worked by scrambling typed messages using a polyalphabetic substitution system; recipients of the message could only decipher the message if they had the exact setting for the rotors of the machine. In 1932 three Polish cryptologists working for the Polish Cipher Bureau first broke the cipher. Just before Germany invaded Poland, the Poles involved British military intelligence, which established the Government Code and Cipher School at Bletchley Park. Because the Germans believed their code was indecipherable, they used ENIGMA machines to communicate military movements. Intelligence breaking high-level military action was codenamed "Ultra" and eventually became the name by which all high-level intelligence was known to the Allies. Winston Churchill would tell King George VI after the war that "it was thanks to Ultra that we won the war."

During World War II the Manhattan Project—the program initiated by Franklin D. Roosevelt in 1939 to develop the first atomic bomb—was so secretive that Roosevelt's vice president, Harry Truman, only found out about the program in 1945 after Roosevelt's death. A Soviet spy ring learned of the project before the FBI knew and by 1944—a full year before Harry Truman would first learn the details of the project and authorize the atomic bombings of Hiroshima and Nagasaki—Klaus Fuchs, a top physicist on the Manhattan Project, had telegraphed important scientific information related to construction of the weapon to the Soviet Union.

As we've moved into the digital world, breaches of sensitive data and systems have accelerated. In our digital world, it only takes one user—one node—to serve as the leak through which the data breaks, then replicates like a virus.

In our vernacular, "data breach" has come to mean unauthorized acquisition of personal information, as opposed to simply security breaches that may or may not involve personal information. This is largely a result of data becoming more personal. At the same time, data breaches are becoming more common—or at least it appears that way, given recent high-profile examples. In 2014, Target CEO Gregg Steinhafel resigned after a thirty-five-year career at one of the largest companies in the world. Steinhafel was at the helm during one of the largest retail cyber-attacks in history. Between November 27 and December 15, 2013—at the height of the holiday shopping season—hackers stole the financial or personal information, stored in Target's systems, of an estimated 110 million consumers. In recent months both Home Depot and JPMorgan Chase have reported unauthorized access to their vast computer networks. Each day seems to bring new revelations of data breaches across a host of diverse corporations.

Data breaches don't always take the form of massive theft. On Monday, April 22, 2013, the Twitter account for the Associated Press was hacked and this tweet was sent out:

The stock markets reacted quickly to the news—under the initial assumption that the news was in fact true. The Dow Jones industrial average dived one hundred points—and recovered almost as quickly once the White House announced the president was fine. Nevertheless, the gate had been opened: hacking just one Twitter feed was enough to destabilize the entire U.S. stock market.

Data breaches are this century's bank robberies. No matter how strong the vault, no barrier is perfect. Perhaps the best we can hope for is that the organizations under assault stay one step ahead of the thieves, although as we saw in the case of Edward Snowden, sometimes it's not even a matter of sophisticated burglary. Sometimes it just takes one person to walk off with the files.

SOCIAL MEDIA BLUNDERS

The very thing that makes social media networks so powerful is exactly what makes them so dangerous. I'm referring of course to Metcalfe's Law. We only hear about the high-profile scandals in the news—the Anthony Weiners, the celebrity Twitter missteps, and the occasional government faux pas. But imagine how many more happen in any given day to regular people: the co-worker who mistakenly maligns his boss over Twitter; the teenager who accidently posts a party picture of underage drinking or drug use; and the people, young and old alike, who have been led to tragic ends by online bullying. Mark Twain famously said, "A lie can travel halfway around the world while the truth is putting on its shoes." If only Twain had lived in the era of social media, he might have said that the lie could get all the way around the world, three times over, before the truth came out.

Social media has destroyed careers and lives. Indeed, there's no point in even trying to list all the incidents of this kind of destruction, so frequently do they occur. Never before in the history of mankind have the thoughts and idle musings of so many been so widely available. It is a phenomenon without precedent. Is there any better word to describe the social media landscape as it exists today than chaos? Nameless people become your "friends," privy to details about your life and your thoughts. Because of the nature of digital, they can do anything to your posts, some of it illegal, but most totally legal.

Facebook has more than a billion users and adds over half a petabyte of data every twenty-four hours, which adds up to about 180 petabytes per year. Twitter has more than 600 million users. In theory the "nodes" on these platforms are not all connected to one another, in that you generally can't see what another user posts without first establishing a direct link with that user—a degree of individual choice that is meant, in theory, to limit unwanted data sharing. But the truth is that there's nothing stopping any one of your followers or friends from sharing what you posted, and on and on and on. And it wouldn't take long for that sharing to spread to everyone. We are all extremely closely knit, as a 2011 study confirmed. Of the 721 million Facebook accounts (and 69 billion friendships among them) at that time, 99.6 percent were connected by six degrees or fewer; Stanley Milgram's longstanding finding was still intact.

Individual social media companies have tried to impose a sense of order on the madness, with differing degrees of success. Facebook, a platform originally designed for a community of college students, is open to anyone, anywhere. In response to Facebook's exponential growth, the company has had to put in place a labyrinth of privacy measures to give the users some semblance of control over the data they post. But no amount of security is going to change the fact that the moment you post anything on Facebook, it can be replicated as easily as pressing a button. Again, it is difficult, if not impossible, to confine data.

POLITICAL UNREST

The Tea Party, Occupy Wall Street, the Arab Spring—every one of these political movements began via digital data. The Tea Party started following a February 9, 2009, television appearance by CNBC's Rick Santelli, whose rant against government bailouts and spending rallied traders on the floor of the Chicago Stock Exchange, who began cheering on live TV.[2] The video quickly went viral and in a matter of weeks, "Tea Party" protests started springing up all over the country. Occupy Wall Street began in the summer of 2011 after the leftist magazine *Adbusters* emailed its subscribers saying "America needs its own Tahrir"—in reference to Tahrir Square in Cairo, Egypt. The email campaign ballooned from there, until thousands of protesters "occupied" Zuccotti Park in Manhattan, with thousands more staging their own "Occupy" events across the U.S. Both movements began with and grew from digital data, which allows the freedom of speech guaranteed in the First Amendment to be exercised in its most immediate and efficient form.

For the most part, the Tea Party and Occupy protests were peaceful movements. Neither has led to any mass uprisings or true political upheaval in the United States, but that's not to say it can't happen. The ability of digital data to allow millions of people to communicate with each other in near-instantaneous fashion is a power never before seen in human history. In our country at least, this power has been used for mostly peaceful ends. But in less democratized nations, where the ruling government is less tolerant of opposition voices and less finicky about repressing such voices, that power has unleashed chaos.

We saw it in the Arab Spring in the winter of 2010, when dissidents across the Arab world organized against their ruling governments via social media. A poll taken a few months after the protests found that nine out of ten Egyptians and Tunisians surveyed said they were using Facebook to organize protests or spread awareness about them. All but one of the protests called for on Facebook ended up coming to life on the streets.[3] Halfway across the world, Americans monitored their Twitter feeds to watch live updates stream in from protesters on the ground, confirming the fact that Twitter is truly a global channel, no matter whom you follow.

Whether these massive protests would have happened without digital data is an open question. What's certain is the role digital data had in starting and growing the protests from a few dozen instigators into hundreds of thousands of participants. Perhaps this doesn't look like "chaos" from the point of view of the protesters, but certainly the governments being protested against rued the day Facebook and Twitter were created. "Chaos" doesn't necessarily have negative implications. Particularly in this case, the chaos unleashed by digital data is simply a period of instability in a certain facet of human life that had previously been stable, or ordered. In many of the countries where the Arab Spring took place, the people had been living under oppressive regimes for decades. Yet now, empowered by digital data, they were able to rise up against their oppressors.

LEGISLATIVE AND REGULATORY UNCERTAINTY

And even if the protest movements unleashed by digital data in the West have been relatively peaceful and non-threatening to the government, that hasn't stopped lawmakers and regulators from tampering with digital data *as a way to assert control*. In 2012, Viviane Reding, at that time the European Union Justice Commissioner, pointed out that there were twenty-seven laws that applied to data in Europe—

many of them dating back more than a decade. A similar situation exists in the United States, where regulations and laws that were drafted for an analog world fail to take into account the chaotic and intensely fast world of digital. While many laws were designed to deal with data irrespective of its origin or type (and some remain useful today), the digitization of large amounts of previously uncaptured information has fundamentally changed the traditional definition of data.

A recent effort in Congress revealed how terribly ineffective so many government attempts to assert control over digital data really are. House Bill H.R. 3261 Stop Online Piracy Act (SOPA) and Senate Bill S.968 Protect IP Act (PIPA) would have made website owners responsible and punishable for how users interact with their sites. My employer, the Consumer Electronics Association, came out strongly against both bills and played a key role in killing these proposals. Their defeat is indicative of how the way things used to be done in Washington is over.

We're going to look at the SOPA-PIPA debate in detail in chapter eleven. But the story in brief is that the content lobby—music studios and Hollywood producers—managed to convince enough lawmakers that SOPA and PIPA were urgently needed solutions to the problems of unauthorized downloading and copyright infringement. The bills' backers hoped to squeeze the legislation through Congress without much fanfare, before anyone actually understood its implications. But once word got out—as of course it would, given the properties of data—the response was swift and effective. CEA played a pivotal role in getting the word out and stopping the legislation before it could pass.

On January 18, 2012, Wikipedia, Google, and thousands of other, smaller websites (including the Consumer Electronics Association) coordinated a service blackout to raise awareness among users. Wikipedia would later say that more than 162 million people viewed its banner. Other protests against SOPA and PIPA included petition drives, with Google stating that it collected over 7 million signatures; boycotts of companies and organizations that supported the legislation; and an opposition rally held in New York City.

All of this was fueled by digital data. We're going to discuss legislative and regulatory issues at greater length in chapter eleven, but for now there are two key takeaways: First, the speed with which the digital world is changing means that regulations and laws can't help but be hopelessly outdated, ineffective, or dangerous soon after passage. Second, current laws, such as the Digital Millennium Copyright Act, which were written and passed during a different era, are also egregiously outdated. Nevertheless, until something better comes along, they are the law of the land.

Today laws and regulations tend to be geographically limited. But data travel is not isolated to just the United States or the Eurozone. Data can travel across country borders as fast it can travel within some geographically defined boundary. What are the business and legal ramifications in the future when data doesn't respect geographically defined borders? You can already feel the friction. In July 2014, a New York judge ruled that U.S. search warrants can reach the digital information of persons stored overseas. If upheld, this ruling means data anywhere could be accessible to American law enforcement. This poses a potential danger to American tech companies that want to provide secure cloud storage to both domestic and foreign clients.

As increasing spheres of our lives are digitized, regulation not intended to have overarching influence will gain more power. Take for example the role that copyright is playing today. As Larry Lessig points out, throughout most of history, copyright played a minor role in how ordinary people engaged with their culture. Copyright regulated a very small amount of our interactions. As Jessica Litman observes,

> At the turn of the century [1900], US copyright law was technical, inconsistent, and difficult to understand, but it didn't apply to very many people or very many things. If one were an author or publisher of books, maps charts, paintings, sculpture, photographs or sheet music, a playwright or producer of plays, or a printer, the copyright law bore on one's business.
>
> Booksellers, piano-roll, and phonograph record publishers, motion picture producers, musicians, scholars, members of Congress, and ordinary consumers could go about their business without ever encountering a copyright problem.

But today, that's no longer the case, as Litman further explains,

> Ninety years later, the US copyright law is even more technical, inconsistent, and difficult to understand; more importantly, it touches everyone and everything. Technology, heedless of law, has developed modes that insert multiple acts of reproduction and transmission—potentially actionable events under the copyright statute—into commonplace daily transactions. Most of us can no longer spend even an hour without colliding with the copyright law....[4]

As Lessig expounds, "In the digital world very few uses are copyright free, because in the digital world, of course, practically all uses of culture trigger copyright, because all uses produce a copy."

In an analog world we don't make a copy when we share a physical copy or sell a physical copy. It's simply passed on. But in a digital world, every single use produces a copy. As Lessig concludes, The U.S. "copyright system is an enormously inefficient system that doesn't even tell us practically who owns what."[5]

Chaos.

ECONOMIC UPHEAVAL

The digital transition has also had a pronounced impact on the economic structure of diverse industries. The global economy is now more than six years into a recovery from the global recession of 2008. However, unemployment remains at elevated levels. The U.S. economy, for example, has actually more than recovered what was lost during the recession in terms of industrial production, but unemployment in the United Sates is still at what we once would have considered "crisis" levels. While some of the job losses in 2008 and 2009 were driven by cyclical factors, there are significant lingering structural issues impacting the recovery in employment, where there is a mismatch between available workers' existing skill sets and jobs. This mismatch is most acute in data-oriented jobs. Research by MGI and McKinsey's Business Technology Office suggests that the U.S. is facing a shortage of 140,000 to 190,000 individuals with analytical expertise and 1.5 million managers and analysts with the skills to understand and make decisions based on the analysis of data, and this estimate could easily be off by multiple factors.

Meanwhile, our lawmakers send up pious odes to the manufacturing industry, as if nothing has changed in the U.S. economy since 1960. President Obama famously blamed "ATM machines" and other digital technologies for the poor job recovery. He wasn't wrong, but it's useless to blame a force you can't stop. The whole trajectory of the global economy is shifting, and as the largest and richest economy in the world, the United States and its Western allies can't act as if we're still working in the analog world.

Numerous factors are involved in the lackluster recovery, but chief among them is simply that the digital world has transformed our economy. Much of what we're seeing is the chaos that results when the available workforce simply doesn't have the skills or knowledge to fill the jobs that are available. Politicians certainly don't help the situation by promising to "save" this industry or that industry.

There are certainly some policies the government should (and shouldn't) adopt to ease our way through this disruptive moment. But for the most part, we're simply going to have to deal with the repercussions of a wave of "creative destruction" unlike anything we've seen since the Industrial Revolution—and we're just getting started. The second-order effects of the digital revolution have economic implications that we aren't yet prepared to address.

THE DIGITAL DIVIDE

I've traveled throughout the world and have seen stark inequalities in socioeconomic status, living conditions, educational and employment opportunities, and access to and use of technology. The digital divide is the nomenclature used to describe inequalities that exist between users of technologies. As we increasingly digitize more around us and rely more heavily on digital technologies, the risks and implications of a digital divide heighten.

The consequences could be severe in those countries whose citizens aren't benefiting from the gains of digital data. As we know, in a digital world, things move very quickly—and that includes wealth creation and accumulation. As the Western world accelerates, the Third World finds itself more and more cut off from the advances of modern technology.

As I've argued, we should expect that digital data will trickle down to these poorer countries. Indeed, there are many nations in Africa and elsewhere where mobile phones are more prevalent than running water. This is all because of the cheap cost of digital products, which will only get cheaper as we move forward. Nevertheless, poorer nations risk becoming stranded, which could only exacerbate political chaos and economic uncertainty.

THE POTENTIAL FOR MISPLACED SENSORIZATION

We generally presume that everything will eventually be brought into the digitized world. In my years attending CES and following the technology industry I've seen a wide array of things brought into the connected digital world that I would never have expected. I've seen espresso machines running operating systems, toilets that connect to your phone via Bluetooth, toothbrushes that sync data, cups that analyze and digitize everything you drink out of them, and sensorized and connected planted pots, cows and other livestock, Christmas trees, and even baby diapers.

This all suggests the question, when everything can be digitized, should it be? One of the great challenges in the road ahead of us is to determine what should be sensorized and connected and what shouldn't be. The more digitized devices we have, the more connected devices we will have. As Apple discovered, this is a good way to clog the system with useless apps. When everything is digitized, might we find ourselves in a world where the chaos never gives way to order?

I firmly believe that market forces will help us solve this question—consumers, after all, will want some order imparted to the choices before them. But it may take some time for the market to adjust—meaning that in the near term we can expect to crawl through a bunch of kitten videos on YouTube before we find what we're looking for.

DIGITAL REMORSE

In my discussion on the properties of data, I discussed how data seeks permanence. This desire for permanence is something that we as both creators and users of digital data must balance with our own personal desires to forget the past.

The interest in forgetting the past creates friction when data seeks permanence. In 2014, the European Union Court of Justice ruled that individuals have the right to request that search engines remove links with personal information about them when the information is "inaccurate, inadequate, irrelevant or excessive"—the so-called "Right to be Forgotten" ruling.[6]

In 2010 in Greenwich, Connecticut, Lorraine Martin was arrested with her two grown sons when a police raid of her home found marijuana, scales, and plastic bags. The case was eventually dropped and her official record expunged when she agreed to take drug-prevention classes. But even with a clean police record, Martin was unable to secure work she was well qualified for. Upon Googling herself, she discovered headlines such as "Mother and Sons Charged with Drug Offenses" in the online new archives. The *New York Times* reports her lawyer saying, "It's essentially a scarlet letter. She's become unemployable in spite of the fact that she has no criminal arrest record."[7]

Google's transparency reports show continued requests from governments and courts to remove online material. Bill Keller writes, "Editors tell me they are increasingly beset by readers who once cooperated with a reporter on a sensitive subject—nudism, anorexia, bullying—and years later find that old story a recurring source of distress. (It's called 'source remorse.')"[8]

The more that is digitized, the greater the potential for digital remorse.

WE ARE ALL DOGS NOW

In 1993, Peter Steiner published a cartoon in the *New Yorker* depicting two technology-savvy canines talking while one of the dogs sat on a chair at a computer. The caption, which read, "On the Internet, nobody knows you're a dog," might have been true in 1993, but by 2015 we have begun to digitize large swaths of information. Not only is it now known that you are a dog on the Internet, but it is also known exactly what kind of dog. When you are online, what you are doing online, and a host of details about your online activities are now all known. As we digitize more and more of our surroundings, it isn't just what we do online that is recorded; it's also what we do anywhere and everywhere. This will bring with it profound benefits, as we'll discuss in the following chapters, but it will also bring with it great challenges that will have to be addressed as digitization increases.

Data's chaos-order cycle plays out every day. I've identified a few points of friction, but these are by no means the only ones. At any point along our path to our digital destiny, you can see chaos giving way to order and order forming from what was once seemingly chaos. The nature of digital is that things will be in a constant state of change. In other words, there will always be eddies of swirling madness amidst the relative ordered calm—just as in our physical universe. The world will never again go through "eras" that last centuries or longer. We will be constantly moving from one paradigm to the next, year after year, month after month. The order in some areas of life will dissolve into chaos, and the chaos in other areas will solidify into order. This cycle, exciting and punishing all at once, will be the new normal, if anything can ever be called "normal" again.

CHAPTER 6

In the Year 2025...

"Where is the information we have lost in data?"

—Hiroshi Inose and J. R. Pierce, *Information Technology and Civilization*

"It will feel like *Minority Report*," said a Microsoft general manager in 2007 about the company's "Surface" technology, which was advertised as turning an ordinary surface into a computer.[1]

"*Minority Report* is one possible outcome," said a chief development office of his company's retina scanning security technology.[2]

"In the film [*Minority Report*], the billboards rely on scanning the person's eyeball, but we are using RFID technology," said a research scientist at IBM's innovation laboratories, regarding a new billboard that will deliver individualized advertisements to each passer-by.[3]

These quotations represent a sort of running joke within the tech community. Since its premiere in 2002, the Steven Spielberg movie *Minority Report,* starring Tom Cruise, has come to stand for a reality-based look at the future. Although it is a science fiction flick based on a novel by Philip K. Dick, *Minority Report* is no *E. T.* or *Star Trek.* Even its principal sci-fi element, the existence of three women with psychic powers (the "precogs"), is a metaphor for real technology's increasing ability to foresee human action.

For readers who haven't seen the movie, here's a quick synopsis. In the year 2054, Washington, D.C., is the center of an experiment known as "precrime." This special unit within the police force uses the psychic powers of three women, the precogs—for pre-cognition— to predict when a crime will take place. When one of the precogs gets a vision, the information is relayed to the crime unit, which is able to identify the crime, the perpetrator, and the victims. Once these have been established, officers swarm in—on jet packs no less—to apprehend the would-be criminal before he or she actually commits the crime. The moral dilemma is apparent. However, precrime seems to work and Washington has been murder-free for six years.

The movie portrays a future remarkably free of the usual sci-fi tropes. There is no "beaming" or "warp drive." There are no aliens. There isn't even any space flight. In fact, one of the most striking things about the future in *Minority Report* is the way it looks so much like our present—at least certain corners of it do. For instance, when Tom Cruise jetpacks his way from the highly advanced downtown Washington to intercept a would-be criminal, he and his team land in a suburban neighborhood that doesn't look much different from today's Washington suburbia.

Nevertheless, the future is different. It's just not strange—with the possible exception of the robotic spiders, but those are only in development.[4] Spielberg's 2054 is possible in all of its elements, minus the precogs. Some of the tools, gizmos and tech are already in our present, or not that far off.

As a 2002 story in *Wired* tells it, Spielberg "convened a think tank" of futurists, biomedical researchers, computer scientists, and even architects. Their goal was to "contribute ideas on the future of cities and suburbs, play and work, nutraceuticals and good old-fashioned grub."[5] Over three days, these folks did nothing but imagine what *will be* based on *what is*. They did such a remarkable job at this that, nearly fifteen years later, techies still seem to explain whatever it is they're developing as "kinda like *Minority Report*."

As I think about the future, I fight the urge, but I too inevitably find myself constantly circling back to the same line: "kinda like *Minority Report*." As in...

...the driverless cars of the future will be "kinda like *Minority Report*."

...the security systems of the future will use biometrics retina scanners "kinda like *Minority Report*."

...advertising in the future will be personalized "kinda like *Minority Report*."

...computing in the future will be an immersive experience, using virtual reality and sensors, "kinda like *Minority Report*."

And so on.

Which is not to say that a movie perfectly pegged our future in every detail. But if Spielberg's vision has any major flaw, it's that he was too generous with the time it would take to see such a future. By 2054, we'll likely be far beyond what *Minority Report* portrayed. Now if the movie had been set in 2024, then perhaps it would have been closer to the mark.

But that's hardly the fault of Spielberg or his assembled think tank. The reality the movie depicts was far from the reality we were living in in 2002, when the film was released. For instance, in 2002 there were no smartphones. There was no 4G network. There were no tablets. Facebook didn't even exist. So we can forgive the film-makers for overshooting the mark by a couple of decades.

With Metcalfe's Law, Moore's Law, the near-ubiquity of comput-ing, and the dirt-cheap prices of sensors, sources of massive change are coming together quickly. So rather than talking about what the year 2054 will look like, let's start with something a bit more imme-diate. Let's start with 2025—a short ten years from now. What will it look like?

Well, kinda like *Minority Report* ...

A WORLD OF INSTANTANEITY AND EFFICIENCY

Before we get to specifics, we need to understand first where digital data is leading us. For that, let's return to two properties of data I mentioned in chapter one—data is both instantaneous and efficient. Recall that data is immediate and therefore seeks instanta-neity; it wants instantaneous recognition and response. Data comes to us slowly only because of the imperfections of human recording and analytical devices. And remember that data is constantly moving toward efficiency, abhorring friction, removing barriers, closing distances, destroying the moments between recognition and response, making us invent better ways of understanding it.

Why does data seek instantaneity in understanding and response? The first and simplest reason is that data itself is instantaneous. From the thought that pops into your head to the drop in temperature outside to the number of cars on the highway, data comes into being instantly and is continuously changing. To be of any use, however, data must be captured and stored, and this creates an immediate divide between when data comes into being and when it can be utilized.

My dad grew up outside of Las Vegas and as a boy once had a job that required him to sit on the side of a lonely highway heading into the Mojave desert and count the number of cars as they drove by. Sitting under whatever shade he could find, my dad would mark a tic in his notebook each time a vehicle passed him and tally the total at the end of the day. This was analog data collection at its finest. Up until the digital era, instantaneous data was captured non-instantly. In other words, there were great big swaths of time between data creation and data capture. Given what we were able to capture, however, the lag didn't mean much to us—we didn't have the competency to do anything with instantaneous data capture anyway. With the start of the digital era and the advent of sensors, we not only have the tools to capture data the instant it is created, we also can utilize the data instantaneously. That same highway into the Mojave desert undoubtedly exists today but it most assuredly is monitored with sensors such as cameras, which are in turn connected to computer systems that analyze data in real time.

That sensors are able to "record live"—and will do it exponentially better in ten years—means that the instantaneous data that was of little to no use to us suddenly becomes extremely useful. The best way to understand a particular piece of data's usefulness within the continuum of time is to grasp that the more time that elapses since its creation, the less actionable or useful the data generally becomes. This is in part because of the constantly mutating nature of data. Despite the fact that we often treat data as discrete, data is continuously changing. Think of memories. Memories are merely the brain's method of capturing and attempting to retain instantaneous moments. But the further removed you are from the moment, the less actionable is the memory. Take a practical example. You visit the doctor because you experienced severe pain in your abdomen the night before. You remember the pain, but relaying the intensity of the pain from memory to the doctor, who will use the data to make a diagnosis, is an imperfect solution. Wouldn't it have been far more useful if the doctor had been by your bed monitoring you while the pain was present? Or better yet: Wouldn't it have been exponentially more useful if a device had been recording the pain level and storing it for your visit to the doctor?

Let's take another practical example. History is replete with different periods and areas marked by massive food shortage problems. I've seen some of this firsthand. Many of these shortages arise because of misallocations of information. Suppliers of food are unaware of shortages and unaware of market prices. I've heard stories of food rotting on farms only miles away from desolate starvation. Often the message just wasn't received.

With technologies like telephones, we shortened the space between data creation and response. With the Internet, we shortened it even more. The more we shorten that gap, the better the outcome. Again, data seeks instantaneous response. But as long as we still need human input to record data or humans to convey data—for example, if someone has to check to see if food resources are getting low or notice that supply and demand are mismatched—we still cannot get from data creation to instant response even with connectivity and smartphones (two elements still in short supply in many parts of the developing world) Can we shorten the time between data creation and response even more? With sensors we can.

The more quickly we can close the gap between action and reaction, the more potent data becomes and the more useful it is to us. Another way to put this is that data constantly moves toward efficiency. In some ways, the efficiency of data is far easier to grasp than its instantaneity, because if we know one thing about technology, it's that it helps make our lives easier—or more efficient.

But with digital, we now have a tool to create a level of efficiency far beyond anything that came before it. Because we will be able to shorten the time from data creation to response, we will remove the barriers, the friction, that otherwise impede our response to data. The food supply story is a good example. Even if we assume the best possible supply chain in remote areas of places like Africa, there are still multiple layers through which the data must travel to elicit a response. In short, the process is inefficient. Digital helps remove those layers, shortening data's transmission chain, skipping potential inhibitors, and triggering a faster response time.

Another example can be found in retail. Digital has already greatly streamlined our shopping experiences. With e-commerce, we can shop without physically entering a store. The data—what we want to buy—is input into the online store's platform and we respond more quickly than is physically possible if we are visiting multiple storefronts in search of what we plan to buy. Digital data is slowly changing everything we know about how we shop. In the fall of 2011, the world's first physical virtual store opened deep beneath Seoul, Korea in the Seolleung station of its subway system. There large displays on the walls depicted an assortment of items, from milk to electronics to detergent. With the scan of a smartphone individuals could purchase items and have them delivered to their home. Recently Amazon was awarded a patent for what it calls "anticipatory shipping." Essentially Amazon wants to start shipping you a package before you've officially clicked "buy." The retailer would predict what you want before you actually buy it by relying on—what else?— digital data. Factors such as previous orders, product searches, historical purchasing behavior, demographics, and wish lists, among any number of other categories of data, could help inform algorithms designed to predict your next purchase before you actually make that purchase.

Financial transactions have long been the poster child for what digitization can mean for data. Financial transactions exhibit many of the properties of data that I outlined in chapter one. Data has been pushing financial transactions and financial markets from the very first trade in the beginning of time through the tulip mania of 1637 and subsequent speculative bubbles to the myriad of financial transactions today. In fact, financial bubbles and ensuing corrections are wonderful examples of how data cycles between chaos and order. The history of financial transactions and financial innovation has frequently been motivated by instantaneity and a push towards efficiency, interspersed with periods of chaos. Digitization has exacerbated this cycle on almost all fronts.

Initially digitization simply moved services that we might have used in the physical world to the online world blossoming before us. Digitization allowed us to check our bank account balances online or initiate transactions that took the customary few days to clear. But this was just the beginning.

Today many on Wall Street rely on fast data networks and complex algorithms to execute rapid trades. This high-frequency trading moves billions of dollars in and out of securities in fractions of seconds based purely on digitally available information—a topic Michael Lewis explores in his most recent book, *Flash Boys*. To date, two major market crashes have been blamed on this data-driven trading method. The first, on May 6, 2010, resulted in a market loss of roughly 9 percent in less than 5 minutes, but most of the loss was regained within the hour. The second flash crash was precipitated by the hoax Associated Press tweet on April 23, 2013, which I discussed in the last chapter. While estimates vary, it is generally believed that high-frequency trading accounts for 50 percent or more of total US equity trading volume.

Wall Street traders aren't the only ones being pushed by digitization. Today individuals can send cash to each other instantaneously and often free of transaction fees through services such as Paypal or Square Cash. While data is infinitely divisible, the division has often been difficult to accomplish in practice, but digital is making it very easy, especially with financial transactions. I can split a lunch with a friend and pay for my share across a digital payment network or split the fare for a service like Uber seamlessly.

The financial transaction component of a service is clearly an area where individuals toggle between digital and physical. While services like Uber are rendered in the physical world, every other aspect of the service, from ordering a driver to payment, is done in the digital world. Prior to the digital revolution that brought us the smartphone and the accompanying digitization of data such as our location, many of these services could not have existed. Today I can buy coffee at Starbucks or a salad at Chop't through their respective mobile apps—bringing digitization to otherwise analog services.

In 2014, Disney introduced MagicBand, which allows guests to check in at FastPass+ entrances, unlock their hotel doors, and also pay for purchases within Disney theme parks. The innovation relies on a short-range radio technology called Near Field Communications (NFC), which is also steadily being built into smartphones and smartwatches. Apple embedded NFC into their iPhone 6 and Apple Watch exclusively for the use of their payment system Apple Pay, launched at the same time.

But what if all of these approaches are just hybrids that are temporarily bridging the gap between the physical and the virtual? Already we've seen the rise of decentralized and completely digital currencies like bitcoin. Today a variety of retailers have begun accepting the new digital currency, and even some politicians are accepting campaign contributions in the form of bitcoin. We are seeing the move towards completely digitized forms of financial transactions. Whether it be through algorithmic stock trades that are completely defined and executed by data or via diverse mobile payment systems that not only bring digitization to physical services but also introduce entirely new services that were impossible without digital devices and digitized information.

By shortening the time it takes from data creation to reaction, by removing the barriers that inhibit data's propensity to flow freely, digital has unlocked a world where the new standard of nearly everything will be: what I want, when I want it.

THE CUSTOMIZING POWER OF DIGITAL

Digitization enables customization, and financial transactions are just one of the areas being impacted. Take for example the rise of 3D printing, which is entering into homes today. If we advance 3D printing by ten years, then we can easily envision an age when many U.S. households own 3D printers and these devices are able to manufacture common everyday products. In this scenario, you buy your product online and the company transmits specifications to your 3D printer. In a few moments, you have your product. Not only will you be able to build entire products essentially from scratch, but you will also be able to completely customize many of the products you buy today. This is all made possible by making the physical digital.

The instantaneity and efficiency of data lead toward one point: a digital experience entirely tailored to your identity. By identity, I mean exactly who we are as individuals—the literal definition. Today we all have multiple "identities"—offline, online, financial, professional, social, consumer, and so forth. These identities, which span our contact with digital data, from our social media presence to what we buy online, are separate at the moment. They are different ways that these platforms view us as users. My bank sees me quite differently than my favorite clothing store sees me. Facebook has a different view of Shawn DuBravac than does my bank. All of these platforms see the real me, they just see me parts of me from different angles and thus can't capture the whole me. These platforms do indeed capture a tremendous amount of data on us and it is growing by the minute. Much of this data is details we didn't even know we were offering, but still, what they know about us is a fraction of what there is to know.

For example, Zappos knows the kind of shoes I've bought in the past and can offer me similar shoes it knows will entice me; but there is much Zappos doesn't know about me. It doesn't know that I'm going on a ski trip in a month. Nor does it know that I plan on moving to a warmer climate in about a year. It also knows scant details about my usage patterns when it comes to all of those shoes I've bought. The best it can do right now is guess, like Amazon's "Recommended" scroll bar, which uses your viewing history and purchase details to provide suggestions. But a future of retailers knowing more about me and subsequently tailoring offerings to meet my needs is coming. The future isn't that far off when everything you buy online—and especially the clothes you buy—will be influenced by hundreds of digital data points. For example, I might authorize one of many wearable fitness devices to transmit my running data to Zappos. In the future I might know down to the step when it is time to replace my running shoes.

As Brynjolfsson and McAfee put it, "We're heading into an era that won't just be different; it will be better, because we'll be able to increase both the variety and the volume of our consumption."[6] But that's only half the story. We are also going to increase fit. Understanding you—the customer—better is the goal of nearly every digital company.

Your devices, from your smartphone to your tablet to your car to your entertainment system, will have near-complete knowledge of you as we move towards 2025. They will know your identity in the literal sense of the word. Some of them will have picked up the data about you in the traditional way—because you've chosen to input it—but most will automatically capture the data, because their primary input system will be sensors and they will be tied into additional sensor systems. You won't have to manually input much, if anything at all. The sensors will capture most of it and then—here's the really interesting part—will be able to share that information across devices and platforms via the Internet of Things. Because all "things" will be connected, they will all be able to exchange data about you, improving their delivery of whatever services they offer.

Let's take my Spotify subscription as an example. Clearly Spotify knows what type of music I like. It knows which songs I listen to most often and which ones I've lost interest in. If Spotify could communicate, it would be able to give you a darn good profile of Shawn's musical tastes—probably better than anyone in my family or my friends. It "knows" my musical identity in a very intimate way. What

Spotify doesn't know is where I like to listen to certain music; and my mood at the time that I need to hear a certain song.

So what if it did know all these things and much more? My preference for a particular song or genre is the data. Finding that song in my library and pressing play is the reaction or response to that song. There is a level of friction, of inefficiency, involved in this process. I must look for the song, then play it. Granted, it's not a tremendous amount of friction. Yet imagine the alternative: What if Spotify knew exactly what I wanted to hear when I wanted to hear it? What if it knew I was out for a run and began playing the songs I wanted to hear while running? What if it knew I was in the car and played the songs I enjoyed while driving? What if it knew I was in a down mood and picked songs to pep me up? What if it knew, when I visited a certain place, that a particular song was associated with a strong memory of that place, and immediately played that song?

Through the use of sensors and the merging of our many "identities" into one holistic identity, activities such as listening to music could change drastically by 2025. The music service of the future will know exactly what we want to listen to given our a) situation, b) environment, and c) mood. Quite an offering, you say! Yes, it will be, but the music service of 2025 won't be performing all these tasks on its own. Rather it will be connected to several other devices and services that perform these tasks much more efficiently, receiving the information acquired via sensors and other digital data like past history, analyzing the data in an instant, and knowing seemingly intuitively (though more accurately algorithmically) what song I want to hear.

This level of communication among devices is possible right now, at least in theory. For example, fitness wearables might know when I'm out for a run. All that's required is for the fitness wearable to be able to communicate with the music service. My smartphone knows from its sensors when I'm driving. Again, being able to communicate this information to the music service is all that is required to supply the perfect driving tunes for me.

We can extend this idea into nearly every corner of our lives, until finally we get to the moment when the sci-fi element of "precogs" in the movie *Minority Report* becomes reality. With machines acquiring billions of bits of data on us every day, it's only a small leap to believe that one day machines will be able to *anticipate* our actions. I hesitate to use the word *predict*, because that involves an absolutely precise knowledge of future actions. A device can anticipate that I will want a certain service at a particular time of the day, week, and so forth. It can't—and likely won't—be able to tell me what I'll be doing this time next year.

Which is not to say that across the population, machines won't be able to collate massive amounts of data to identify things like crime "hot spots" or predict a rise in a certain type of crime. Indeed, research from Matthew Gerber at the Predictive Technology Laboratory at the University of Virginia has found that publicly available tweets tagged with GPS coordinates improved predictions for nineteen of twenty-five types of crimes that happened in metropolitan Chicago over a certain period. In ten years, "predictive analysis" crime intervention could be the new normal for law enforcement.

The point is that in 2025 electronic devices connected to the Internet and equipped with powerful sensors will be ubiquitous, surrounding us at all times, acquiring and analyzing data, not only to give us what we need the moment we need it, but to acquire more information on our true identity continuously. Algorithms utilizing this data will constantly be improving themselves to better understand and anticipate our actions.

In short, our digital destiny means that there will no longer be a wall separating our offline and online selves. Nearly every action we perform will be captured, analyzed, and filtered into its appropriate place online. The purpose, of course, is to further close the gap between thought and action, need and fulfillment. The devices and services we use will become extensions of ourselves—our thoughts and actions. Humanity's progression in communication—in recreating the brain's instantaneous signals—will come that much closer to perfection. It won't be perfect. I don't want to give the impression technological progression ends in 2025. But we are on the cusp of global individualization as digital is allowing humanity to personalize our digital experience for one, rather than for many.

THE DIGITAL LANDSCAPE

We can now look at some broad areas where our digital destiny will have the biggest impact. In choosing the topics that follow, my goal was to encapsulate as much as possible across the widest breadth of society. Clearly there are areas of the human experience left off this list, but if you remember data's propensity for instantaneity and efficiency, and understand that the end result is always focused on each of us, as individuals, you shouldn't find it terribly difficult to imagine what those areas might look like in ten years. What follows isn't a rundown of really cool gadgets; rather, my purpose is to explain how certain sectors of society will likely perform in ten years. Trying to predict specific gadgets is a losing proposition anyway—we can more reasonably anticipate how those gadgets will perform.

TRAVEL

As we began the book, so we begin this list. One of the biggest and most consequential innovations on the horizon will be the advent and mass adoption of driverless cars. Even in ten years, the prospect of a completely driverless world is unlikely. Driverless cars will make their first appearance in the cities. From that point, more and more localities will adopt them, in time choosing to make the full transition from drivers to driverless vehicles.

But what does a driverless world look like? For starters, it will be a world in which motor vehicle fatalities will be rare. Given that there are thirty thousand motor vehicle deaths in the United States every year, and millions around the world, this by itself will fundamentally alter how we experience travel. The necessity of licenses, road laws, and police will diminish until they are negligible factors. These were all analog tools created to form order and manage chaos. With our roads safer, police departments will be able to divert their resources to other, more pressing law enforcement matters.

Meanwhile, with driverless cars, there will be a significant, if not total, decline of traffic and gridlock. Cars will move in unison with each other. As the human element is removed, so too are many of the barriers between data and decision. With this very annoying element of daily life eradicated, automobile travel will undergo a rebirth in popularity. Now that cars will operate like trains, more and more people will choose the comfort of their own cars to the mass transit of most urban centers. Trains may even become a thing of the past. "Working from car" might become common terminology as it becomes a common element of life once we are no longer required to operate the vehicle.

There's a tendency to think that humanity is moving toward ever more isolation. Because of the Internet, we don't need to interact with each other face to face anymore. This is a fallacy. Human interaction is a primal need of our species. It's the annoyances—the friction—of traveling to interact that impedes us from pursuing it as we want. Driverless cars are but one answer to that problem. People will be more mobile in 2025, not less.

HOME

As we attempt to envision the home of 2025, perhaps the first thing to get out of the way is the idea of "Internet access." In 2025, homes will be automatically and holistically "connected," either through a centralized computer system hub or through a myriad of decentralized devices, each connecting to what we know as the Internet today. Every electronic device in the home—yes, every one—will be connected to the Internet, constantly. More, many of the objects in the home that aren't currently electronics devices will see sensors embedded in them, through which they will become digitally available. Acquiring data on human activity from the sensors embedded in these devices and objects, computer systems either locally in the home or via cloud networks will analyze petabytes of information in real time. Just as your smartphone now collects and makes available data across an assortment of apps, our homes will function as one colossal smartphone. PCs as discrete devices will be supplemented with computing functionality throughout the home, from its utility systems to the surfaces in each and every room.

Your appliances will be connected, allowing them to pull in relevant data and in many cases operate autonomously. Your refrigerator will know when a certain food item is running low and what meal would be best to prepare tonight; your shower will adjust the water temperature perfectly based on your own body metrics and the surrounding environment; your entertainment system including the TV and the stereo will be entirely personalized, ready for any action at the sound of a command. Your closet, holding your clothes, will know the day of the week, whether you have work, the weather outside, and all of the other factors you now use to decide what to wear.

ENTERTAINMENT

Say good-bye to linear programming as we know it today through network or channel television. With services such as Netflix and the burgeoning industry of on-demand streaming video, the need for television networks with a set and rigid daily schedule of shows will be obsolete. Aside from a few live programs such as sporting events, all television programming will be held in the cloud, ready to stream to your screen of choice.

The programming itself will also be a far more immersive experience than is offered by even the best 3D sets today, offering a 360 degree view that also incorporates your broader senses. Even television shows will be radically different. If you remember the "choose your own adventure" book series of the 1980s, that's the kind of immersive television we might see in ten years. But rather than explicitly deciding the course of a show based on choices you make for the characters, data captured on your mood or level of engagement might implicitly choose for you. Again, we need to remember the personalized nature of our digital destiny. It's "your" show.

Video gaming will reach a new level with augmented reality (AR) and virtual reality (VR). We'll probably still be a while away from the "holodeck" of *Star Trek* fame—where the crew of the *Enterprise* could create nearly any environment they wished, down to the clothes they were wearing, and act in the new reality they had created. But already today there are AR gaming prototypes that bring the game to your home in the same way VR puts you in the game.

It's entirely reasonable that by 2025 the player himself will be part of the game—as in, you won't be controlling your digital person on the screen as he battles his way to the princess; you will be battling your way to the princess.

HEALTHCARE

Digital offers a myriad of possibilities to improve the delivery and quality of our healthcare. Nanotechnology and the rise of wearable tech will realize its fullest potential in medicine. No longer will we feel the need to go to annual check-ups with our physician. Instead, doctors will be able to monitor our health via the devices in and on our bodies. Constantly acquiring and delivering vital health information to our doctors, these devices will serve as our regular doctor visits. We will become active participants in our medical care. Only in the event that something is askew or warrants personal scrutiny will we need to visit the doctor. Corporations also have a huge vested interest in healthcare and related costs. Will we see businesses deploy sensors to aid employees in measuring and monitoring things like posture and activity throughout the day?

The U.S. healthcare system wastes an estimated $700 billion in unnecessary services, inefficient care delivery, and excess administrative costs. We can save much of that money through the promise of digital. When we only go to the doctor when we need to, when the doctor performs only the procedures that she must, when routine maintenance is assumed by automated devices—which will continually monitor our heart rate, cholesterol, blood pressure, and the many other metrics that now require a human being to measure—then we can finally eliminate a great deal of the extravagant costs of our healthcare.

INDUSTRY

Reaching our digital destiny won't come without human costs. Although we'll discuss digital's economic effects in greater detail in chapter thirteen, the very idea of "manufacturing"—and the need for a manufacturing workforce—will be drastically different by 2025. Major industries, from electronics to energy to automobiles to textiles, have already begun the transition to a machine-influenced workforce. This necessarily means that many millions of workers around the world will see their jobs replaced by machines. This is inevitable, no matter how many promises politicians make. And it will be difficult.

The good news is that consumers will experience a fully personalized economy. Products, up till now mass-produced, will be produced specifically for the individual who buys them. I mentioned 3D printers earlier, but before 3D printers progress to the point of mass adoption, factories will be devoted to manufacturing items exactly as the buyer specifies. We already see this mass customization taking hold. Today individuals can go to Puma or Nike or Converse and customize a pair of shoes exactly as they want them—picking the type of material, the shoe sole, where the logo goes, and even the color of the thread used. A human worker could never hope to achieve such a level of specification and mass production, but with digital it is made entirely possible—and entirely affordable.

We cannot overstate the magnitude of this change. In pre-economic times, human beings built things small scale, mostly for themselves or for a small community. We learned, however, that mass production and individualization of labor led to greater productivity and more affordable products. But we lost the personal nature of those products. No longer do we build a chair specifically for us. Of the mass-produced chairs on the market, we buy the one that's best for us. Digital overthrows this millennia-long trend in human economic activity. In 2025, we will be able to mass produce customized items.

NOW FOR THE BAD NEWS

The vision of our future that I outlined is the end of what will be a very long, contentious process. Even though it is our destiny, and thus cannot be avoided, there will be a complex process before it comes into existence. And that will involve many problems—some technical, some ethical, and some political. Our digital destiny will destroy as much as it creates; but habits, jobs, and customs some thousands of years old won't go away quietly. Indeed, some will resist our digital destiny violently.

The remainder of this book will explore the road that lies before us. We will look at each of the areas identified above in a deeper way and ask: How do we get there from here? Make no mistake, we will get there. It just won't be easy or pretty.

CHAPTER 7

Driverless Cars and the New Digital Age of Travel

"These car robots don't look like something from The Jetsons; *the driverless features on these cars are a bunch of sensors, wires, and software. This technology 'works.'"*

—Tyler Cowen

On January 15, 2009, at 3:27 p.m., a U.S. Airways Airbus A320-200 struck a flock of geese roughly three minutes after takeoff from New York LaGuardia Airport. With 150 passengers on board, Flight 1549 suddenly lost engine power and was in serious trouble. Less than four minutes later, the airplane glided down in the middle of the Hudson River in a splash of water. Just minutes after that, all passengers and crew were successfully rescued. There were no fatalities and only minor injuries.

The captain of Flight 1549 has since become a national hero. Chesley Sullenberger III, or "Sully" as he is now known to a grateful nation, was the latest in a distinguished line of heroic American pilots who have accomplished the extraordinary in moments of extreme danger. Sullenberger's decision to ditch the plane in the Hudson, rather than make for a nearby landing strip, likely saved everyone on board. His handling of the plane was also top-notch, as a post-crash investigation proved. In his account of the crash, *Fly By Wire: The Geese, the Glide, the Miracle on the Hudson*, journalist William Langewiesche notes how the National Transportation Safety Board attempted to mimic Flight 1549 detail for detail: "A simulator was programmed to duplicate the circumstances of Sullenberger's bird strike… and four pilots were enlisted to fly a series of attempts on LaGuardia…. In every case where the pilots were allowed to respond immediately to the loss of thrust by making a quick turn back to the airport, every one of them was able to land safely…. In recognition [of the need to account for reaction time] the NTSB then imposed a thirty-second delay before allowing the simulator pilots to fly their returns—and every one of them crashed."

For this reason, it's not wrong to label what happened that cold Thursday morning as a "miracle." I recall a commercial pilot telling me once that they practice emergency water landings during their annually required simulator training but they never expect to have to do one and they certainly don't expect it to actually be successful. Had a lesser pilot than Captain Sullenberger been on the stick, it's likely we would remember the crash as "the Tragedy on the Hudson." Langewiesche does not want to discount Sully's role in averting disaster. "It was a beautiful piece of flying," writes the author, a pilot himself.

But Langewiesche argues that the attention on Sullenberger has obscured the role that the airplane itself played in saving everyone on board. This isn't a surprise. After all, an airplane can't very well appear on the morning talk show circuit or write a bestselling book.

But it's no disservice to Captain Sullenberger to point out that he couldn't have done what he did without the digital "fly-by-wire" control system in the Airbus A320.

To greatly simplify a very complex piece of equipment, the "fly-by-wire" system uses gyroscopes fitted with sensors that are mounted in the aircraft to sense movement changes along the pitch, roll, and yaw axes. Any movement results in signals to the computer, which automatically moves control actuators to stabilize the aircraft. The term for this is "flight envelope protections," which limit how badly a pilot can screw up mid-flight. For instance, a pilot can't suddenly plunge the airplane into the ground. The "fly-by-wire" system won't allow it. The system intervenes should pilots accidently attempt to exceed the aerodynamic and structural limits of the aircraft. For Flight 1549, the system freed Captain Sullenberger to focus on pressing matters—such as where to land the plane—while the computer took care of most everything else.

But "fly-by-wire" alone doesn't promise an infallible flight. Five months after Sully's miraculous landing on the Hudson, Air France Flight 447, plunged into the Atlantic Ocean en route to Paris from Rio de Janeiro. The crash was mysterious precisely because the Airbus 330 had the same digital "fly-by-wire" system. It was the first crash of an Airbus 330 in commercial passenger service in the fifteen years since the introduction of the fleet; the 330 was largely considered foolproof because of the fly-by-wire system.

Sifting through the black box data, the investigators discovered that as the plane passed through the high reaches of a large storm, ice crystals formed on the plane's pitot tubes (sensors that measure air speed) subsequently deactivating the autopilot system and eliminating stall protection (because the fly-by-wire system needs airspeed data to fully function). With the captain resting, the less experienced co-pilot made a series of fatal errors that led to the plane's stalling out and crashing into the ocean. A little over a minute passed before the pitot tubes unfroze, but even with a correct airspeed registering and stall warnings blaring in the cockpit, the junior pilot failed to put all of the evidence together and make the appropriate adjustments. Because the fly-by-wire system overrides pilot commands that would induce a stall, one hypothesis suggests the junior pilot might have ignored the sounding stall alarms believing that stalling the plane was impossible. But with the autopilot off there were far fewer limitations—or "flight envelope protections"—on the pilot's control over the aircraft.[1] Tragedy ensued.

The two very different outcomes of these two flights have a lot to say about the future of sensorized and subsequently digitized autonomous vehicles. On the one hand, we see how digital greatly restricts the scope for human error. Indeed, human and computer complement each other. A computer could not have landed on the Hudson by itself, while Captain Sullenberger likely could not have landed on the Hudson without the computer. On the other hand, the benefits of technology handling the continuum of complexities produce a reliance on automation that creates its own complexities should automation fail. Ice crystals form, and human beings make mistakes. Earl Wiener, who was a professor at the University of Miami, NASA scientist, and pioneer in human-automation interaction outlined some of the intricacies of automation in "Wiener's Laws," part of which I quote here:

17. Every device creates its own opportunity for human error.

18. Exotic devices create exotic problems.

19. Digital devices tune out small errors while creating opportunities for large errors....

23. Invention is the mother of necessity....

27. It takes an airplane to bring out the worst in a pilot.

28. Any pilot who can be replaced by a computer should be.

29. Whenever you solve a problem you usually create one. You can only hope that the one you created is less critical than the one you eliminated.[2]

While some of these principles are specific to air flight, they apply broadly to our dance with technology.

AUTONOMOUS VEHICLES ALREADY SURROUND US

These incidents also demonstrate another fact about autonomous vehicles, one largely overlooked by the public: how heavily we already rely on autonomous and semi-autonomous vehicles every day. Aviation is just one industry in which digital sensors have revolutionized aircraft operation over several decades. The first electrical systems were tested by the Soviets in the 1930s. By the 1960s, the U.S. government was using "fly-by-wire" systems on NASA spacecraft, such as the Apollo Lunar Landing Research Vehicle, which astronauts including Neil Armstrong used to practice landing on the surface of the Moon.

From there, "fly-by-wire" systems moved into military aircraft. In 1972, the Navy's carrier-based F8-Crusader was the first aircraft to use a digital "fly-by-wire" system.[3] The Airbus 320 model that Capt. Sullenberger flew that fateful morning was first outfitted with a digital "fly-by-wire" system back in the 1980s. Indeed, it was the first commercial aircraft to use "fly-by-wire" with no mechanical backup.[4]

Today, digital "fly-by-wire" systems far more advanced than the systems used by the NASA astronauts or Navy pilots decades earlier come standard in many commercial aircraft. While it's a stretch to say that the computer flies the plane itself, there are indeed only a very few human-required controls remaining on airplanes these days. I'll admit that this fact does remove some of the romance we have associated with derring-do pilots from Charles Lindbergh to Chesley Sullenberger. Nevertheless, the digitization of flight is one of the biggest factors contributing to the overall safety of airline travel.

As a *New York Times* article noted in 2013, the trajectory of airline crashes has marched steadily downward since the dawn of the jet age. While "[i]n 1985, more than 2,000 people died in dozens of crashes," the *Times* pointed out, 2012 "was the safest since 1945, with 23 deadly accidents and 475 fatalities, according to the Aviation Safety Network, an accident researcher. That was less than half the 1,147 deaths, in 42 crashes, in 2000."[5] According to statistics from the Bureau of Aircraft Accidents Archives, 2014 saw the fewest crashes in a single year since 1926.[6] Of course one major difference between then and now is the presence of a digital "fly-by-wire" system on nearly every commercial aircraft today. Today's commercial airlines rely on thousands of sensors collecting up to a half terabyte of data on each and every flight.

Aviation isn't the only industry implementing autonomous elements and relying on digital data. Today, farmers are using GPS data together with sensor data collected from a farm combine's grain flow and grain moisture sensors to measure yield variability and farm productivity. Farmers can then utilize these data to vary inputs such as water or fertilizer based on the expected productivity of a given field. In 2011 Kinze Manufacturing introduced the first autonomous planting system, and today these autonomous agriculture systems are being utilized in the operation of working farms.

The value in autonomous tractors, and other agricultural machinery, isn't so much safety as it is efficiency. A self-driving tractor removes the need to hire a driver. Relying on digital data ensures that fields are planted in the most efficient manner while avoiding waste through things like double planting. Harvesting crops is a demanding round-the-clock exercise, and autonomous equipment doesn't need to stop to eat or sleep. The net result is a higher yield with far fewer resources expended or wasted.

In short, autonomous vehicles already surround you—whether you're getting on a flight or buying produce at the supermarket. Sensor data and other digitized information was initially used to improve safety performance, but it is now being utilized for a host of additional applications such as increasing productivity. Each of these applications brings with it a host of complex obstacles and difficulties. As you can see, the progress in tackling any one particular set of problems for one given industry doesn't necessarily transfer to another. Each industry has its own unique challenges, which determine not just the speed of development but also the kind of technology that will be implemented. But at the root of all of these solutions lies digitized data.

THE OPEN ROAD NO MORE

It is curious that autonomizing travel began with flying since far more people have died on the roads than in the air. You'd think that the pursuit of a safer, autonomous road vehicle would have been a necessity long ago, spurring innovation at least as early as the innovation that gave us autonomous aircraft. Why aren't our cars at least as automated as our airplanes? It's an interesting question, and one we don't necessarily have to answer in full. Part of the explanation is in the nature of driving itself. For starters, driving a car requires far less skill than flying an airplane. Because nearly everyone can drive, the pressure since the beginning of the automotive industry hasn't been so much safety as cost: How can we make cars more affordable tomorrow than they are today?

That downward push in car prices precludes the sort of upfront expense that would be required to innovate driverless features. Indeed, one of the factors keeping driverless cars off the road is the prohibitive cost of the range of autonomous systems needed to guide the car safely. As we saw in chapter 4, we can all but guarantee that those costs will decrease over time. Even so, it will be quite a while before driverless cars are, like Ford's Model Ts, affordable to the workers in the plant. For airplanes, the single biggest factor has always been safety: protecting the pilot (and the passengers). Cost, while important, has been a secondary consideration.

A second explanation is that a car is a necessity for a vast majority of people—which makes for a very competitive market, so that R&D departments don't have the luxury of straying too far afield from the basics. As the industry matured and the car become available to middle- and lower-class people, innovation focused on how to differentiate—by performance, mileage, comfort, and recreation—

what were in essence very similar products. For many years, consumers were much more concerned about the number of cup holders a vehicle had than about many other features the vehicle could have offered. Diverting resources away from those features to experiment with driverless cars didn't enter the equation until very recently.

But now digitization, by generating efficiency and lowering costs, has enabled car manufacturers to turn their attention to driverless features such as adaptive cruise control, lane assist, and parking, in a pattern of innovation and diffusion that we will see play out numerous times as we move more deeply into a digitized world driven by data. The automotive industry began by solving discrete, well-defined problems: How should the speed of a vehicle adjust when a vehicle has cruise control engaged and is approaching from behind another vehicle traveling at a slower speed? How should a vehicle respond when a driver veers outside of his lanes? How can a vehicle aid a driver in parallel parking?

All of these problems and others like them—all extremely difficult problems even in isolation—are being slowly solved. Adaptive cruise control adjusts a vehicle's speed in order to maintain a safe distance from a vehicle ahead. Lane departure warning systems (or "lane assist") alert drivers if the vehicle leaves its lane—and increasingly can take countermeasures to help keep the vehicle in its lane. "Parking assist" systems can steer vehicles into parking spaces unaided by driver input. Each of these solutions uses a variety of sensors to digitize the physical space the car is in and then translates this newly digitized data into actionable events.

From here, these independent solutions will be combined. It's only a matter of time before the whole vehicle becomes fully autonomous.

A third reason we are only now starting to see driverless cars on the road is that we operate vehicles in complex systems. Within a given day, the average vehicle will confront a large number of unexpected obstacles—from pedestrians to balls rolling into the street. The skies contain far fewer unexpected impediments. Autonomous systems of any kind hunger for digital data, and in the case of driverless cars, we needed really reliable (digital!) mapping and other infrastructure improvements before cars can drive themselves.

The final explanation for the fact that our driverless future has not already arrived has nothing to do with digital and everything to do with analog. The car has always occupied a symbolic role, particularly in American culture. We associate driving with, in a word, freedom. Going out on the open road is emblematic of the way settlers traveled West earlier in American history. The frontier is gone, but Americans still have their car, which they can take almost anywhere at any time. Teenagers count the days until their sixteenth birthday, because a license equals freedom. The stereotypical middle-aged man buys his red Corvette—not for utility, but because it's a symbol of success, happiness, and freedom. Inseparable from the car's association with freedom is our ability to operate it manually. *We* drive the car; *we* have control.

The first two problems are easily overcome. Technological progress has made a focus on driverless features economically viable. The third problem is being overcome now through the deployment of yet more sensors and the digitization and communication of a tremendous amount of data. In the end, the fact that driverless cars will save tens of thousands of lives will be decisive. But it will take some time, decades in fact, to resolve the fourth road block to our driverless future. It will be tremendously difficult to disassociate cars from independence, particularly among older generations. As automotive travel becomes more restricted with the removal of the human element, older drivers might recoil at the loss of manual operation. The

car as they've come to understand it—with all its symbolic over-
tones—will cease to exist. We should expect that this issue will affect
public adoption of driverless cars in the short term—as it has already
affected the sales of autonomous tractors.[7] A powerful piece of
Americana will become a thing of the past, just like the frontier.

At the same time, autonomous vehicles will free us from long
commutes filled with "wasted" time stuck in traffic. We will be able
to dedicate travel time to other recreational activities such as reading
or sleeping. Certainly autonomous vehicles will bring about cultural
changes, but they will also bring about other significant changes.
Autonomous vehicles will forever change every aspect of the in-
vehicle experience, from the interior of the car to what we do inside
while traveling. Driving in the future will look nothing like it does
today.

HOW SOON?

In the spring of 2014, expectations for driverless cars reached a
fever pitch after Google presented a prototype of a driverless car with
no steering wheel or pedals. Suddenly everybody was wondering
when these new wonders of the road would actually be on the road.
Google itself set expectations at about five years in the future
(roughly 2020) for its autonomous vehicle to be available. But even
if Google hits this rather early date, we're still a long way from living
in a world where driverless cars make up the majority of cars on the
road, much less all the cars on the road.

To better understand what the future holds we need to ditch the
idea that cars will become fully autonomous overnight. The technology just isn't quite there yet—which also means that an autonomous
vehicle, were one to even exist, would be cost-prohibitive. For instance,

one of Google's earlier models, which mounted sensors on an existing car, had about \$150,000 in equipment.[8] That's only about \$30,000 less than the median home price in America.[9]

As with most technological advances, the transition from full human responsibility to full vehicle responsibility will be gradual. You could say that some features found already on some car models—such as Chevrolet's "Super Cruise" and Ford's "Active Park Assist" for automated parallel parking—demonstrate that the transition has already begun. To get a better understanding of how this transition will play out, the National Highway Traffic Safety Administration (NHTSA) has helpfully defined five different "levels." As the NHTSA defines them:

- **No-Automation (Level 0):** The driver is in complete and sole control of the primary vehicle controls— brake, steering, throttle, and motive power—at all times.
- **Function-Specific Automation (Level 1):** Automation at this level involves one or more specific control functions. Examples include electronic stability control or pre-charged brakes, where the vehicle automatically assists with braking to enable the driver to regain control of the vehicle or stop faster than possible by acting alone.
- **Combined Function Automation (Level 2):** This level involves automation of at least two primary control functions designed to work in unison to relieve the driver of control of those functions. An example of combined functions enabling a Level 2 system is adaptive cruise control in combination with lane centering.

- **Limited Self-Driving Automation (Level 3):** Vehicles at this level of automation enable the driver to cede full control of all safety-critical functions under certain traffic or environmental conditions and in those conditions to rely heavily on the vehicle to monitor for changes in those conditions requiring transition back to driver control. The driver is expected to be available for occasional control, but with sufficiently comfortable transition time. The Google car is an example of limited self-driving automation.
- **Full Self-Driving Automation (Level 4):** The vehicle is designed to perform all safety-critical driving functions and monitor roadway conditions for an entire trip. Such a design anticipates that the driver will provide destination or navigation input, but is not expected to be available for control at any time during the trip. This includes both occupied and unoccupied vehicles.[10]

If you're keeping track, today we're between Level 0 and Level 1. The transportation consultancy firm Fehr & Peers predicts that driverless cars won't reach 50 percent of all highway vehicle miles traveled until 2040. It won't be until about 2070 that that number approaches 100 percent.[11]

But Fehr & Peers's is just one of many predictions out there. Even if its predictions are accurate, companies such as Audi, BMW, Cadillac, Ford, GM, Mercedes-Benz, Nissan, Toyota, Volkswagen, and Volvo have begun testing vehicles at all other Levels up to and including Level 4. Audi predicts that it will have its A8 model with "driverless technology" on the market by 2018. BMW, Volvo, and Nissan say they will have driverless lines by 2020; Ford, by 2025.[12]

There's still the issue of price, of course. Even when these cars hit the market, it's highly unlikely their price points will be within the reach of average consumers. (We've already mentioned Google's $150,000 hybrid model.) We can assume that the early adopters of driverless cars will be the rich; much as the early adopters of the first cars were the upper crust of post-Victorian society.

It would be a mistake, however, to assume that all driverless cars will be a model unto themselves—at least in the early phases. What's far likelier is that manufacturers will offer "add-ons" that will make an existing model anywhere from semi-autonomous to fully autonomous. Put another way, it won't be much different from shopping for a car today and debating whether to get the on-board computer with Navistar technology or opt for the simpler design. So the appropriate way to think of the cost of driverless cars is *how much more* it will cost to make a model autonomous. We are already seeing autonomous-like features show up in cars today.

A 2013 report on the future market of driverless cars from the Eno Center for Transportation analyzed several studies on the topic of price. It found that the first phase of driverless vehicles will have added costs that fall between $25,000 and $50,000 per vehicle. For instance, if you're looking at a non-autonomous vehicle that costs $20,000, you'll have to assume at least a price two to three times higher to make it autonomous. That's certainly in the price range of the wealthiest Americans, but far from realistic for the rest of us.[13]

Yet as we've seen with other digital technologies, particularly sensors, you can expect the price to fall from the effects of Moore's Law and mass production. The Eno Center says that ten years after the first wave of autonomous vehicles hits the market, the add-on costs will fall to around $10,000. If we use the predictions from above, we're somewhere around 2030. Still a ways into the future, but we're getting closer.

The study notes that, according to Erik Coelingh, Volvo senior engineer for AV capabilities, the preferred price increase for the add-ons (meaning, the price that would make autonomous vehicles affordable for anyone) is $3,000. When will this happen? The Eno study says,

> Electric vehicle costs have been declining by 6 percent to 8 percent annually, suggesting that it may be 15 years at 8 percent annual cost reduction to go from $10,000 [autonomous vehicle] mark-up (perhaps possible in five to seven years' time after initial introduction) to a $3,000 mark-up (20 to 22 years after introduction). For comparison, as of February 2013, adding all available driver-assist features, adaptive cruise control, safety options (including night vision with pedestrian detection), and the full "technology package" increases a BMW 528i sedan's purchase price by $12,450, from a base MSRP of $47,800. While these features provide guidance and a degree of automation for certain functions, full control remains with the human driver."[14]

In short, we're still around Fehr & Peers's 2040 prediction as a reasonable point at which we can expect that driverless cars will be affordable to the average American. Thanks to sensors and the digitization of data, it's not a question of *if* driverless cars will replace manually operated ones, but *when*.

BUILDING A DRIVERLESS INFRASTRUCTURE

Yet neither cost nor technology is the largest obstacle from realizing our driverless future. Those things will sort themselves out in time. No, the two biggest obstacles to our driverless future are entirely human in nature. Let's deal with the less obtrusive of the two: infrastructure.

Automobile manufacturers are developing driverless cars for today's roadways, which—apart from traffic signals and other law enforcement tools in urban environments—remain largely unchanged since the days of the Roman Empire. Our roads may be straighter; they may be composed of better, cheaper, and safer materials; and they may form a network more elaborate and extensive than anything those great ancient road builders could have ever imagined. Nevertheless, the basic design of roads has not changed for thousands of years.

Yet if the sensors on driverless cars will be able to interact with other driverless vehicles, communicating, in a sense, for better maneuverability and safety, then why should our roadways not advance along similar lines? Indeed, it seems rather bizarre to leave the whole work behind the travel experience to the cars alone, especially when the technology exists to add roads into the communicating matrix.

That indeed is what should happen—which is not to say that it *needs* to happen. Manufacturers can't predict what innovations the public sector will put on roadways to make them sensorized and compatible with driverless cars, so they're developing the cars under the assumption that the roads won't be sensorized.

But many of the challenges currently facing automobile manufacturers could be overcome more quickly if they could include sensorized roads into their designs. Take traffic signals, for instance. Most traffic signals operate on timers, giving a calculated amount of time to each direction of traffic. In many cities, these timers vary depending on the time of day and are calculated to account for lighter and heavier traffic volumes. More advanced traffic signals employ sensors underneath the road directly before the intersection to trigger a change in the signal. All well and good—except this kind of technology is older than the first Macintosh.

So much more is available, as Los Angeles—a notoriously congested city—has shown. In 2013, L.A. became the first city in the world to synchronize all of its traffic lights—all 4,400 of them.[15] In theory, what this means is that an L.A. motorist can drive from one end of the city to the other and make every green light, just so long as the driver is going the posted speed limit. Of course in reality this almost never happens, except in the dead of night—which hardly helps daytime commuters. The reason it never happens: traffic volume and unforeseen interference. For instance, if you get stuck waiting to turn right as pedestrians cross the road, then you're not going to make every green.

Still, even if the change isn't perceptible to L.A. drivers, the improvement in congestion has been significant. According to the city's Department of Transportation, the synchronized traffic lights have reduced traffic time on major L.A. corridors by about 12 percent—put another way, if a certain route used to take you an hour, it now takes you 53 minutes.

Replacing today's cars with driverless vehicles will only further improve the success of traffic light synchronization. When you remove the human element, which is the cause of so many backups and accidents, and account for the way in which driverless cars will maintain constant speed and specific distances from other cars, you can begin to see how synchronized traffic signals could improve overall congestion. In the end, though, this is a fairly minor step forward and still does not reach the full capabilities and benefits of the technology. Synchronization, for instance, doesn't take into account larger traffic volumes in one part of the city versus another. It treats them all the same way. More advanced traffic light systems will use sensors to determine changing traffic flow and adjust the timing of traffic lights accordingly, not just during rush hour but throughout the day in real time. In 2012, researchers at Carnegie Mellon University began testing an adaptive traffic control "smart signal" system named Scalable URban TRAffic Control (SURTRAC). SURTRAC measures traffic flow and facilitates communication between traffic signals. Early tests have found travel times cut by over 25 percent with wait times cut by 40 percent. Smart Signals and other traffic-sensing technologies are being deployed today in major urban areas from Singapore to Chicago.

For data to make real impact, it must be communicated across different objects. Synchronizing traffic lights is one thing, but getting them to exchange information within a network of related objects is the real goal. In 2014, the U.S. Department of Transportation (DOT) and the National Highway Traffic Safety Administration (NHTSA) continued to advance vehicle-to-vehicle (V2V) communications systems that will enable cars to share relevant information such as speed and traffic flow and even potentially the presence of pedestrians (by recognizing their mobile phone's wireless signal). The recently concluded Safety Pilot Model Deployment in Ann Arbor, Michigan, demonstrated proof of concept and collected valuable data showing the potential of V2V.

By mid-2015, DOT's Research and Innovative Technology Administration (RITA) anticipates releasing vehicle-to-infrastructure (V2I) deployment guidance. V2I enables the widest sharing of digitized information. V2V together with V2I will set the foundation for autonomous transportation. Infrastructure sensors will help drivers avoid trouble areas, such as accidents, wet or icy surfaces, and construction. The mapping system on many of our GPS devices and phones tries to do some of this for us—if with less than stellar results. Services like Google's Waze rely on crowdsourcing user-inputted information. A far better way to determine road conditions would be to have the roads themselves tell us—or our cars. With a driverless car, we wouldn't even know what was happening, since the car would automatically alter its route to the destination, with zero human input.

Roadside sensors also would greatly improve the design of road networks, by giving city planners massive amounts of data in real-time. Cities collect this information already, but doing so digitally would add a level of accuracy and efficiency unlike anything they've ever experienced. The result would be more efficient networks, less congestion, and safer roads. Potholes, fallen trees, downed power lines, flooding—all of this data would be immediately captured by the roadside sensors and sent to the city's transportation department. I can't say that the powers that be would fix the problem any quicker, but at least you could be assured they were aware of it.

Roads also can be constructed—and maintained—with driverless cars in mind. As Clifford Winston, a senior fellow at the Brookings Institution, has shown, there are a number of innovations we can make to the way roads are constructed that would better serve a driverless market. For instance, "Most highways in major metropolitan areas operate under congested conditions during much of the day. Yet highways are designed around standards based on higher free-flow travel speeds that call for wider but fewer lanes. Driverless cars don't need the same wide lanes, which would allow highway authorities to reconfigure roads to allow travel speeds to be raised during peak travel periods. All that is needed would be illuminated lane dividers that can increase the number of lanes available. Driverless cars could take advantage of the extra lane capacity to reduce congestion and delays."[16]

Winston also advises that future highway construction should account for the differences in weight between cars and trucks. Right now, highways are built so that all lanes can accommodate both. But driverless cars would benefit if they had their own lanes, with trucks given their own lanes. Such a separation would allow the more numerous and narrower car lanes to be built with only a few inches of pavement; the truck lanes would be built with about a foot of pavement. As Winston notes, this would be a win for both taxpayers and driverless cars: "Building highways that separate cars and trucks by directing them to lanes with the appropriate thickness would save taxpayers a bundle. It would also favor the technology of driverless cars because they would not have to distinguish between cars and trucks and to adjust speeds and positions accordingly."[17]

Winston goes on to argue that governments need to be focused on the coming driverless future when they determine transportation expenditures. For instance, he criticizes the Obama administration's focus on high-speed rail at the expense of better, more innovative highways. It's not that there isn't a future in high-speed rail; it's more that whatever future there is will pale in comparison to our driverless-car future. Given the scarcity of transportation resources—a headache for any municipality—and the meager 0.5 percent of its budget the Federal Highway Administration spends on R&D, the various government agencies involved in U.S. roadways need to prioritize spending with an eye to the future.

WHAT IS (AND SHOULD BE) LEGAL?

As for the second great obstacle to our driverless future, I'll give you a hint: it starts with a "g" and ends with a "t."

It's no indictment of any politician, party, or government to say that today's road and vehicular laws did not anticipate driverless cars or trucks. But neither do any federal or state laws prohibit them, technically speaking. Stanford's Center for Internet and Society analyzed current federal regulations, legal frameworks from all fifty states, and the 1949 Geneva Convention on Road Traffic to determine the legality of driverless cars. The short answer is that autonomous vehicles are "probably" legal under current laws. As the study lays out,

> The Geneva Convention, to which the United States is a party, probably does not prohibit automated driving. The treaty promotes road safety by establishing uniform rules, one of which requires every vehicle or combination thereof to have a driver who is "at all times…able to control" it. However, this requirement is likely satisfied if a human is able to intervene in the automated vehicle's operation.
>
> N[ational] H[ighway] T[raffic] S[afety] A[dministration]'s regulations, which include the Federal Motor Vehicle Safety Standards to which new vehicles must be certified, do not generally prohibit or uniquely burden automated vehicles, with the possible exception of one rule regarding emergency flashers.
>
> State vehicle codes probably do not prohibit—but may complicate—automated driving. These codes assume the presence of licensed human drivers who are able to exercise human judgment, and particular rules may functionally require that presence.[18]

This is good news as far as it goes, but far from completely satisfactory or sufficient. The absence of prohibition generally makes something legal, but in our world it's safe to cover your bases. Four states (Nevada, Florida, California, and Michigan) and Washington, D.C., have already passed legislation specifically allowing autonomous vehicles, with varying levels of regulatory oversight. Colorado, on the other hand, rejected a bill that would have permitted autonomous vehicles in 2013. Abroad, the United Kingdom has permitted testing of autonomous vehicles on public roads.

While the four trailblazing states and D.C. certainly deserve kudos for being ahead of the curve, it's also true that we're not quite at a point where laws are necessary. Since we can't really say what form autonomous vehicles will take, it's a bit premature to start imposing regulations on them. For instance, the California and Michigan statutes, while leaving most of the regulatory concerns to their respective departments of transportation, require that a licensed driver be behind the wheel at all times. This stipulation assumes that autonomous vehicles will have both a wheel and a manual override system. It's likely the first models will, but after that, who knows?

Nevertheless, it's certainly not premature to start discussing these issues, so that when the day comes that autonomous vehicles are ready for the market, there are as few legal and regulatory hurdles to overcome as possible. The closer driverless cars get to the market, the more manufacturers will be looking to state and federal regulators for guidance on what is allowable. It would greatly expedite what could become a laborious and controversial debate to start the conversation today.

With that in mind, here are the most important areas for exploration by state and federal regulators and politicians, for which I turned again to the Eno Center study:[19]

Autonomous Vehicle Licensing: Car manufacturing is fast becoming less an industrial endeavor and more a tech endeavor. Production of lighter-weight, more fuel-efficient, and more digitally compatible models has started to tilt the balance away from "car factories" as we've come to know them. The major car companies understand this—the city of Detroit certainly understands this. If Google's entry into the automotive industry wasn't the first danger sign, then Google's appointment of Alan Mulally, former CEO of Ford Motor Company, to its Board of Directors[20] should have left no doubt. With Silicon Valley now in the car business, the question becomes: What defines a car manufacturer?

It's really a question for the regulators, as the Eno study makes clear. In Nevada, for example, to be a licensed autonomous vehicle manufacturer, a company has to have "a minimum of 10,000 autonomously driven miles and documentation of vehicle operations in complex situations," meaning on the vehicles' ability to obey signs and recognize crosswalks and school zones, "the presence of pedestrians, cyclists, animals, and rocks, and recognition of speed limit variations," and so forth.[21]

We can certainly expect new entrants into the automotive world in the coming years, in addition to the traditional incumbent companies, and not just because of digitization. Unlike some of the industries we'll meet later in this book, the automotive industry has been quite ready to accept change—indeed, to lead the innovation itself. Starting with electric vehicles and now with autonomous vehicles, the industry is not content to sit idly by (pun intended). This foresight will give the traditional automotive industry a very long—if not indefinite—lease on life, particularly compared to some other incumbent businesses that fight against the digitization of their respective industries. Instead of ignoring the digital tsunami headed their way, the car manufacturers decided that they had better learn how to ride it.

Litigation and Liability: Perhaps no other legal matter is more shrouded in uncertainty than the question of liability when it comes to autonomous vehicles. No autonomous vehicle will be 100 percent safe because the cost would be prohibitive. There will be accidents on the driverless roads of our future, and there will be fatalities. But by removing the human element from the equation, we're suddenly in a liability no-man's-land. Autonomous vehicles open up large ethical questions. The most commonly cited scenario is what happens when a pedestrian or an animal jumps in front of a driverless car and the vehicle cannot respond quickly enough? Should the vehicle be programmed to protect the occupants (and therefore hit the pedestrian or animal)? Or does the car try to protect the pedestrian or animal at the cost of potentially injuring someone inside the vehicle?

For a manually operated car, this liability question is perhaps easier to answer. Human beings respond to these situations instinctively and aren't held responsible for what are pure accidents. We might veer out of the way and consequently injure ourselves or we might not and in the process injure someone outside of the vehicle.

For driverless cars, the response will likely be preprogrammed into their DNA. The software to take into account digitized data and respond accordingly is written long before the accident. To swerve or not to swerve was a question asked and answered long before the vehicle took the road. These are complex ethical questions that are difficult to resolve, and the resulting liability questions are equally complex. Autonomous vehicles will be programmed to take in a variety of digitized information—the speed of the vehicle, the road conditions, and the presence of other objects or people nearby. But ultimately the autonomous vehicle will have its destiny prepro-grammed. Who's liable in an accident that involves injury remains uncertain.

These are questions that manufacturers will need to answer in concert with regulators—with a healthy dose of advice from legal counsel. I have no doubt that a set of standards will arise, but so far we haven't even started the conversation.

Security: As with everything else that has gone (and will go) digital, concerns over security will play a large role in the coming debates. When roadways are digitized via sensors, controlled by a central server, you have created a prime target for hackers and cyber-terrorists. We already rightly worry about the security of our other infrastructure systems—air traffic control, electrical, Internet, and so forth—and we would be remiss not to give the security of our roadways serious attention.

To give you a taste of what might be possible, the Eno study notes the following scenario: "a two-stage computer virus could be pro-grammed to first disseminate a dormant program across vehicles over a week-long period, infecting virtually the entire U.S. [AV] fleet, and then cause all in-use AVs to simultaneously speed up to 70 mph and veer left."[22] You can imagine the ensuing horror from such a devious plot. Carjacking becomes car hacking.

As we already know from other widespread attacks, successful and not, this sort of possibility is a new fact of life in our digital age. Fear should not keep us from pursuing a driverless future, but fear should act as an impetus to make that future as secure as possible. Vehicle manufacturers are well aware of the risk and taking them into account as they develop autonomous vehicles of the future.

Privacy: As with security, privacy concerns are a staple of our new digital world—and no less with driverless cars than with mobile phones or other digital devices. As we'll discuss in more depth in a future chapter, your driverless car will have a ton of information about you and it will communicate that information to other devices in your home, office, and other locations. It's likely that your car already collects a lot of information, since 96 percent of those sold in the U.S. have event data recorders that investigators can access in the event of a crash. While these recorders capture just a bare fraction of what the driverless cars of the future will record, you can see that the process has already begun—and probably without you even knowing it!

The Eno study lists five questions that are of paramount importance when it comes to the data your car will collect:

1. Who should own or control the vehicle's data?
2. What types of data will be stored?
3. With whom will these data sets be shared?
4. In what ways will such data be made available?
5. And, for what ends will they be used?[23]

You can probably guess the types of data your car might collect: where you go, when you go, and how you go. You can easily list a number of third parties who would be interested in this type of data, such as government and law enforcement agencies, insurance companies, advertisers, and thieves. The law surrounding who has access to this data is developing.

There remain many unanswered or only partially answered questions. At the same time, not all data sharing is bad. Would you voluntarily give up the data on the route you take to work if you knew that city officials would use that information to improve road networks, cutting down on traffic? Would you give an advertiser access to some of your data if it led to discounts and other cost savings?

Questions about data sharing and privacy always involve trade-offs.

WHERE THE RUBBER MEETS THE ROAD

Autonomous vehicles offer a wide array of wonderful benefits. The NHTSA estimates that of the 5.5 million accidents in the U.S. each year, 93 percent have a human as the primary cause.[24] Every year there are 2 million injuries from motor accidents, with thirty thousand fatalities. The annual economic cost of all this mayhem on the road is $300 billion, or 2 percent of GDP.[25] In short, automobile accidents generate a tremendous amount of pain and suffering, in terms of lives and dollars. What's more, these are the *best* statistics we have ever had. And globally, the numbers are far worse. Automotive safety, combining better technology and laws, has improved exponentially since Ralph Nader first made the seatbelt a national priority. But as long as imperfect, fallible human beings operate cars, trucks, and motorcycles, there is a floor beneath which we cannot pass.

That is why driverless cars will become our future, regardless of the obstacles in the way. As opposed to some of the other digital areas we'll discuss in later chapters, there will be few voices of opposition to autonomous vehicles. The cost in blood and treasure is just too high. Not to say that there won't be attempts to prevent or slow the transition. In addition to the obstacles listed above, there are certain special interest groups that will not condone the advent of autonomous vehicles. Autonomous vehicles will displace some of today's workers, and this will likely draw a fight from those quarters, to stymie the adoption of this technology and the growth that will come with it. But I suspect these will be short-lived fights: the promise of driverless vehicles is just too great.

Beyond the safety benefits, there will be other societal repercussions from our driverless future. Less congestion in urban areas will give some urban residents a reason to move outside the city (to take advantage of the now more pleasant or productive drive from the suburbs) and encourage others to move to what were once more congested locations. Density of urban areas will increase while some people will self-select to even longer commutes because the time can now be used more productively. Many of us live where we do simply because of our commutes. In Washington, which I call home, it's common. But with the removal of congestion—not completely, since there will likely always be traffic, but as a major factor—people will have more freedom to live where they want, rather than live where they have to. As you may have noticed, customization is an underlying theme throughout this book. Driverless cars will allow greater customization of where people live.

Another change we likely will see is that people will travel more. Imagine a train that can take you door to door anywhere in the country? If such a train existed, wouldn't you use it more often—as much as, if not more than, airplanes? That's what driverless cars promise. You will be able to go to sleep in Chicago and wake up in Florida, your car having driven the whole way while you slumbered peacefully. As I said earlier, there's a popular perception that our future will be less social and less mobile than yesterday. Driverless cars disprove this. We will have more reason to travel—anywhere at any time—because we will have removed the single greatest barrier to travel: hassle. When you can work, read, and sleep while a car takes care of the travel headaches, then we've entered a new world.

Even planes, our modern world's greatest innovation in travel, won't be able to provide the comfort and luxury of your personal driverless car. They might be faster, but they don't have the freedom to leave whenever *you* are ready to leave.

So what we might lose in the old fashioned notion of the "open highway," we gain in the ability to remove the barriers that have made getting from Point A to Point B so bothersome and so dangerous.

CHAPTER 8

The Internet of Me!

"Getting information off the Internet is like taking a drink from a fire hydrant."

—attributed to Mitchell Kapor

"[The Internet of Everything] will be five to ten times more impactful than the whole Internet revolution has been so far."

—John Chambers (Cisco) 2014 CES keynote address

At the 2014 International CES I noticed a product that was quite unique. It seemed to encapsulate the whole history of digitized data in one convenient item. Was this, I wondered, the product that would revolutionize the home and finally make good on all the promises and expectations of the much-vaunted Internet of Things?

Considering that the product in question was a crock-pot, the answer was obvious: unlikely.

While we might say there are already plenty of digital crock-pots on the market today, they are only digital in the sense that they have a digital interface; otherwise, they are very analog, everyday crock-pots. This one being showcased at CES, however, allowed remote monitoring and control via one's smartphone. The accompanying app told the user the temperature and the elapsed cooking time and allowed adjustments. This new crock-pot is literally part of the so-called Internet of Things. It is digital. It is "connected." It is sensorized.

My tongue-in-cheek praise should not be interpreted as veiled skepticism about this particular product—or of any digital device to come. At the same CES, I also saw Kolibree's smart toothbrush, a digital toothbrush that can track how long you brush, what teeth you clean well, and which teeth could use a little more attention. Revolutionary? Perhaps. The Internet of Things has to start somewhere, and we should applaud those companies who are taking the early risks.

Indeed, these items are just the first of what will be a horde of digital items coming to store shelves—or online outlets—near you. The day will come when not only will you own a digital crock-pot, but the crock-pot will "communicate" with your driverless car to make sure dinner is piping hot the moment you enter the house. As you're getting ready for bed, still full from your amazingly delicious (and perfectly cooked) crock-pot meal, your digital toothbrush will let you know that you've neglected your back molars the last two times brushing and that they have an excess of plaque.

Or not. Maybe digital crock-pots and toothbrushes will never catch on. While you'll certainly be living in a house full of connected items, it's still guesswork which ones will be digitized and online and which ones won't. The experimentation taking place today is most importantly helping to determine which devices and use case scenarios make sense to digitize and connect and which ones don't.

What we do know is that by 2020, according to the IDC "Digital Universe" report, there will be around 30 billion connected "things."[1] Your home, the focus of this chapter, will be connected—not in the way we think of connected homes today, with a cord running to a central modem or even through Wi-Fi connecting a few key computing products—but through hundreds of everyday objects, running independently but communicating with each other.

The purpose of all these connected things—if we can identify a single unifying theme—will be to transform the home into an extension of yourself, or your family. The home, via its devices, will assume many of the daily tasks, annoying chores, and myriad other activities we now perform manually. In the home of the future you will even be able to create the very products you buy online.

A NEW INTERNET FOR AN ALL-DIGITAL AGE

Before we get too far ahead of ourselves, we need to take a step back to where this process began. Of course trying to pinpoint when the Internet "began" is a bit tricky. For the purposes of this chapter, I will define the beginning as when the masses not only knew about the Internet but also had started to use the Internet on a daily basis. That takes us back to 1995.

On October 24, 1995, the Federal Networking Council (FNC), a now defunct creation of Congress and several U.S. government departments, passed a resolution defining the term "Internet." The resolution read, in part,

> "Internet" refers to the global information system that—
> (i) is logically linked together by a globally unique address

space based on the Internet Protocol (IP) or its subsequent extensions/follow-ons; (ii) is able to support communications using the Transmission Control Protocol/Internet Protocol (TCP/IP) suite or its subsequent extensions/follow-ons, and/or other IP-compatible protocols; and (iii) provides, uses or makes accessible, either publicly or privately, high level services layered on the communications and related infrastructure described herein.

It's doubtful whether anyone other than tech historians or industry insiders remembers such an event. That's partly because the FNC resolution didn't change much, other than square away some bureaucratic language. Another reason is that by 1995 the Internet already had gone mainstream—the government might feel the need to define the very word, but the rest of us just wanted to get online. The Internet had reached a stage where it was no longer serving a niche community of scholars, government agencies, and tech geeks. It was now serving customers—or "surfers" in the parlance of the time. By mainstream, I mean that there were some 40 million users online globally—including 25 million in the U.S. Not an insignificant figure, to be sure, but just so we keep things in perspective, today Facebook has more than a billion users.

Nevertheless, 1995 was one of those moments in history where great minds are able to see far beyond the rest of humanity—to the annoyance of the rest of us, who can't believe we missed it. Here are just some other notable happenings in 1995:

- January 18: The venture founded in 1994 by Jerry Yang and David Filo as a directory of other websites is renamed Yahoo! and the Yahoo.com domain is first registered on this date. The company Yahoo! would be incorporated on March 1, 1995.

- March 25: The first "Wiki," created in 1994, is launched on this date in 1995 by Ward Cunningham, whose invention would become the now-ubiquitous Wikipedia.
- May 23: JavaScript, which would become the dominant computing language of the Internet, is released to the public.
- Summer: Larry Page and Sergey Brin meet at Stanford as graduate students. By the end of the year, they will be working together on a project known as "Backrub," which two years later will become Google.
- July 16: Cadabra.com, founded by Jeff Bezos in 1994, goes live to the world as Amazon.com.
- August 9: Netscape goes public, nearly setting a record for first-day appreciation. The extremely successful IPO gives birth to the term "Netscape Moment" and ushers in the dot-com boom.
- August 16: Microsoft releases Internet Explorer 1.0 as an alternative to the dominant browser of the day, Netscape Navigator.
- August 24: Microsoft releases Windows 95—an instant success worldwide. Not only will Windows 95 cement Microsoft's dominance of the OS space for years to come, many of the features still on Windows today (such as the Start and TaskBar) were introduced with Windows 95.
- September 3: Pierre Omidyar launches AuctionWeb.com, the precursor to eBay.com.

- November 22: Disney's Pixar Studios releases *Toy Story*, the first all-digital (computerized) feature-length movie.

If nothing else, I bet I just made you feel really old. But you get the idea: 1995 was a big year. It saw the start of the dot-com explosion, the beginning of e-commerce, the first browser war, and the earliest formulation of an idea that would revolutionize the Internet: search.

These components, expanded and refined though they were over the years, dominated what we would know as the Internet for the next fifteen years. When people said they were "online," we knew exactly what they meant. You sat down at your desktop or laptop, opened a browser, and there you were: online. Using search, you visited certain sites where you could read the news, indulge your interests, and shop. It still sounds remarkably familiar, doesn't it?

That's because we are still largely in this first Internet phase. Forget Web 2.0 or some other jargon. For nearly two decades, the Internet has been what it started to become in 1995; for many of us, it is the only Internet we know. But we are at the tail end of this first phase. The Internet as we've always known it to be is about to become something else entirely.

If you need convincing, answer a simple question: When was the last time you were online?

Not so long ago the answer to this question was straightforward. You were online the last time you were at your computer. Not anymore. Maybe you're reading this on your e-reader—in which case you may be online right now.

Is your smartphone near you? Is it on? Then it's likely online. Does that mean you're online? An email just came through to your phone. You opened it up. Are you now online?

You get the picture. We've entered a grey area with respect to what it means to be online. Our devices, from smartphones to tablets to e-readers, have allowed the Web to creep closer and closer to what was once our offline selves. We no longer "connect" to the Internet as those of us who remember dial-up service did back in the 1990s. High-speed broadband access, which increasingly is wireless and therefore essentially omnipresent, continually connects your devices—and thus you—to the Web.

In other words, the line demarcating online and offline is getting wider and fuzzier.

A few years ago I planted two small blackberry bushes in my backyard. Blackberry bushes propagate quickly. They tolerate poor soils, and the branches root as they reach the ground, creating what is essentially another blackberry bush. Today I have a sweeping and voluminous array of blackberry bushes that still continue to grow out of each other. What was once two small discrete plants is today a massive, rambling, circuitous labyrinth. In the same way, digital data is blossoming and growing and taking root in multitudinous places. We've known it as the Internet, but it is really much more. And it is taking over every nook and cranny of our lives.

As it spreads out, our understanding of the Internet—what it is and what it means to be "online"—will have to change. Indeed, it has already changed, although you might not have seen the transformation in quite these terms.

The change began to occur when we first detached ourselves from the computer screen. In 2014, the number of smartphones surpassed the number of computers in use for the first time. This shift to mobile computing has had a pronounced impact on the way we use the Internet, and more fundamentally on the Internet that we use. And yet this shift is just the beginning. The connected devices that we will see emerge over the next decade will make today's Internet almost completely unrecognizable.

The Internet experience we enjoyed on computers was a common and shared web experience. It was a Mass Produced Web: the same information for everybody. Our Internet experience was also largely linear. We spent time moving among a few select web sites, consuming content made for the masses one page at a time. Aside from your email and tracking a few stock tickers relevant to you, there was no truly personalized information on the Internet. Nor was there much of a "holistic" Internet; the Web was, for all practical purposes, a content medium.

And that's all the mobile Web was when it first appeared. The mobile Web experience began with a general push to drive the Internet as we knew it on the PC onto our mobile devices. Most of the first mobile websites were focused on the mobile browser experience. Beyond a few alterations, it was the same big Internet experience we were having on the PC, simply pushed to mobile devices.

Then things started to change. Companies considered "mobile first" started to appear. These companies took a decidedly different approach when it came to the Internet on mobile devices. At about the same time, companies opened their mobile apps platforms to third-party developers. These developers realized that consumers didn't want just a traditional Internet experience on the go; they wanted a new Internet experience that aligned with their mobile needs.

The personal nature of the mobile phone empowered companies to create a more personal Internet experience. For instance, as opposed to browsing the full Web, consumers wanted bite-sized pieces of the Web—a fast-food version, if you will—catered specifically to them as individuals. Services such as Yelp, which provided users with nuggets of Internet information, started also pulling in physical information from the user, such as GPS coordinates, to provide a web experience that became decidedly personal. The "content-only" nature of the Internet began to change—it now knew where you were in the physical universe.

So rather than plugging in city and state, then your neighborhood, then your restaurant preferences when you're looking for a place to eat, you can allow geo-locating to eliminate the need for everything except the last bit of information. Even that step can be eliminated if you've already let Yelp know which cuisines you prefer.

As I said, this is only the beginning. Today we find ourselves at the precipice of a great Internet revolution. The mobile Web offers just a hint of what this new Internet will look like, as smartphones give just a hint of what connected devices will do. As I've mentioned before, your smartphone is collecting data on you nearly all the time, from the places you visit to the apps you download to how you interact with those apps. Your smartphone knows you quite well. Because of this, the mobile Web—or the version you access via your smartphone—knows you quite well, too. We've moved beyond the realm of browsers—most of our access to Web data on mobile devices comes via apps today. Through these apps you are "online"—and yet information is presented to you in a way that is unique to you. When you pull up your Flixster app, you don't have to tell it where you are in the world. It already knows, and it's already compiled a list of nearby theaters. Likewise, when you pull up your Facebook app to post a selfie with your friends, it's likely that the app already knows who's in the picture. The digital tools for our future are honing their accuracy. We are in a great period of experimentation as we move into the next digital era.

Now imagine tens of billions of smartphone-like objects surrounding us. I use the term object in place of device because we'll see the Internet all around us in common household products. Already we are seeing the Internet appear on our wrist, in our garden, in our bathroom, family room, and kitchen. These Internet-enabled objects will soon be in your bedroom, your garage, and your refrigerator. They will be collecting enormous amounts of data on you in order to provide you with a more streamlined experience, just as a smartphone does. A big difference, however, is that your involvement in this online experience will be trimmed significantly—these things will operate independently of whether you're actively "online" or not.

For now, we are treating the Internet on these new objects just as we treated the Internet when it first moved to mobile devices—as a smaller, more personalized web experience. Wearables, such as the Jawbone Up, require an accompanying app. There are thermostats, like Nest, that you can control with your smartphone. With the aforementioned toothbrush, you can track, via your smartphone, how you brush your teeth.

The connected objects we see today are generally only useful when the data they capture comes to us via the smartphone—or, really, any computer. That's because they don't yet run independently. They still require human involvement to give them any usefulness. As with the invention of writing thousands of years ago, with computers and the Internet human beings were able to record their thoughts and actions in a new medium. It was a revolution in data and communication. But it still required human involvement in the form of data input—a laborious process that greatly limited the new medium's reach. Only with the invention of sensors have humans overcome this age-old problem of manual input. Now a machine captures the data for us.

As with the written word, so with connected objects. Sensors will set them free, creating in the process a unique, individualized web—one running and operating entirely independent of our direct involvement. Imagine a thermostat that not only knows the outside temperature, but also that your internal body temperature rises at night and falls in the morning, and adjusts itself accordingly. You don't do a thing: the thermostat itself, in conjunction with other objects that know your internal body temperature, does everything for you. What's more, these devices will continually capture information about us and our surroundings. While this might scare a few people, the goal is a living experience more tailored to the individual than anything we could have imagined ten, twenty years ago.

Importantly, the whole idea of being "online" will fade away—only to be become a quaint phrase from the early days of the Internet, a phase that lasted between roughly 1995 and 2015. In its place, there will just be the present—no online, no offline, just a seamless transfer of data, response, and actions between you and your connected objects. And at this point many might starting saying, "This feels a little like *Minority Report* …"

THE EVOLUTION OF THE INTERNET

If you want a better look at what this new Internet will become, you should follow what some of the largest tech giants are doing. Take Google, for instance. Brin and Page built a multi-billion-dollar business on the back of a search engine. Google entered the market in the late 1990s just when the most popular browsers, Internet Explorer and Netscape, were battling each other for dominance. Google took no part in this battle. Choosing, instead, to let the big boys fight it out while it attracted users with its simple design and differentiated search algorithms.

At that time, a differentiated search engine was just what the Internet needed. The amount and diversity of sites were both exploding exponentially. It was difficult to find your way around except with a well-programmed search engine. In other words, the Internet had unleashed chaos—in the form of too much information—and Google provided the order. Once more, we see data create order from chaos. Google brought order to the chaotic vastness that was the 1990s/2000s Internet.

Hardly anyone remembers now, but in the early days of Google, a typical search would yield a few thousand hits. To get just a few thousand hits on any search these days, you need to be ultra-specific. Your usual search will return millions, sometimes hundreds of millions of hits. There's still a good chance you'll find what you're looking for, but even so the sheer scale of the Internet seems to render the simple search function a bit old fashioned. That's because it is. Search is a carryover from the old Internet, when the action of searching for something using rather generic keywords would yield relevant results. And it's not as though the Internet is shrinking.

Chaos is taking over once again as the old way of establishing order—search—is becoming insufficient. What's going to happen when tens of billions of objects come online?

This evolutionary change is already under way. As Mark Cuban has written,

> It recently dawned on me that I do far more searches away from Google than I do on Google. I'm not talking about a factor of 2 to 1 or 3 to 1. My off Google searches outnumber Google searches by at least 10 to 1 on a daily or weekly

basis.... If I want to know if anything noteworthy happened in an NBA game, the last place I would search is Google. I would search twitter first. If I want to know if anything interesting happened at an event, the last place I would search is Google. I would search Instagram. The list goes on.... That's not to say I don't use Google...Google is still an important part of our lives.... What I am saying is that I place a significant value on recency for many of my business and personal related searches. Google does a very poor job of indexing and presenting real-time, near-time or even recent information. Which in turn begs the question of whether this lack of recency will impact our ability to trust Google or other search engine results? Or will we just learn where to use Google and where not to use it?[2]

Not only is what we search for changing, but how we search is changing. The frequency of needing to conduct an Internet search with generic key words these days is changing. We are relying more heavily on natural language recognition through voice services like Siri. We've begun asking the Web questions and expecting answers rather than asking for suggestions and receiving recommendations. For the occasional generalized search, we usually end up at sites like Wikipedia. Algorithms have gotten better. The top presented search results receive more than a third of the traffic. The point is, the whole search concept that Google pioneered in the late 1990s is already changing. At the same time, we are overwhelmed by more digital choice than ever before because newer and better ways of bringing order to the chaos are being created just as traditional search fades.

There are the content aggregators—the Wikipedias and Yelps of the world—that collect vast amounts of content in certain niches. There are also apps, which remove the chaos of the browser all together. You're "online" but you're only managing one topic, task, or activity at a time. When you want to know the score of the game, you don't Google it—although you could. You open your ESPN app. Then there are social media sites such as Facebook, Pinterest, Twitter, Instagram, and so forth, which are all helping create perhaps the most significant quality of the Next Internet—the one I call the Internet of Me.

Social networks are the Internet's attempt to impose further order on the chaos. Through these networks we carve out our own little niche of followers, friends, interests, and so forth. So absorbed are we by these little online worlds we've created that they consume much of our online time.

In December 2013, Pew Research found that 73 percent of adults are on at least one social network—and 71 percent are on Facebook. Moreover, 63 percent of Facebook users log in at least once a day, while 40 percent do so multiple times a day. Of Instagram users, 57 percent visit the site at least once a day and 35 percent multiple times per day; and 46 percent of Twitter users are daily visitors, with 29 percent visiting multiple times per day.[3] This is what it means to be "online" nowadays: checking in on our little social networks, which are created entirely for our personal pleasure. Each Facebook page, Twitter feed, and Instagram feed is entirely unique; there are no two exactly alike because there are no two people exactly alike.

If we look at the progression here, it looks something like this:

Surfing → Too Many Choices → Search → Ranking → Too Many Choices → Curation through Apps → Too Many Choices → Social Circles → Too Many Choices → Algorithms.

With each step forward we minimize the chaos a little bit more and expand our customization a little bit more. The notion of a wild, untamed Internet was very exciting when we first began contemplating the possibilities; but we discovered that wild and untamed is a bit of hassle. We'd rather have the efficiency, range, and power that the Internet provides, but customized, ordered, and fitted to our specific identity.

This is why you don't see new search engines trying to rival Google, as you once saw new browsers vying for control. Google itself has also moved beyond search, working to develop and promote a personalized Internet experience. So are Apple, Microsoft, and the other big tech firms. Search was intended to distill the big broad Internet down to something relevant and manageable. Google today delivers customized search results based on your search habits—pulling in personalized data in hopes of creating a more tailored experience. But we are also now seeing digital information distilled in an even more focused way across multiple dimensions. Digitally available information is not only becoming more granular, but also adding in other elements such as time and location. The traditional search function of distilling information has morphed into services such as Google Now, which proactively delivers information that it predicts you will want based on a number of factors including your search habits.

The trend is clear. There will likely always be something of the free "online" Internet as we know it today, where you use a browser, type in a search, and off you go, "surfing" along through hyperlinks and webpages. But that's not going to be your primary day-to-day interaction with the Internet in the future. If "search" defined the old Internet, then the New Internet will be defined by *you*.

A DAY IN THE LIFE ...

Now, what does all this talk about the future of the Internet have to do with the home? Good question. First, like the Old Internet, the New Internet will definitely be experienced in the home. It makes sense that the place where you spend most of your time will be the place where the Internet of Things—connected objects with sensors—will be highly operational. Second, the trend we see the Internet taking is the same trend that will play out in the home: a progression toward near-universal customization. In their book *The New Digital Age: Transforming Nations, Businesses, and Our Lives*, authors Eric Schmidt and Jared Cohen write of this future, "You'll be able to customize your devices—indeed, much of the technology around you—to fit your needs, so that your environment reflects your preferences."

That's all the Internet of Things really is—yet another step toward removing the distance and the friction between data and action. Connected objects capture and analyze the data in a quantity and at a speed that human beings could never match—and then they will act on it, removing the human element from the equation.

So what will this look like in practice? Well, let's create a day in the life of our future selves ...

6:42 a.m.: Your alarm goes off. Is it strange that your alarm went off at the forty-second minute and not the fortieth minute? Not if your alarm knows your sleep cycles and wakes you during your lightest moment of sleep, a much more natural way to wake up than being shocked awake when you might be in deep REM sleep. Gently, the ambient lighting in your bedroom turns on, brightening at the same rate that your eyes can adjust. At the same moment, your shower turns on as well—adjusting the water temperature to match your personal preference, which it has learned. After you shower, your calendar matches the day's events with options in your wardrobe, prompting you with different options and predicting your preferences among the choices presented.

7:30 a.m.: Your coffee is fresh and waiting for you, brewing based upon past behavior and sensor data from throughout your home. You arrive in the kitchen and are prompted with several breakfast choices. The option with the highest rank uses the last of the strawberries because your refrigerator sensed that they were about to go bad. As you eat, you glance through the day's news on one of several screens around you, your favorite news sites arranged in the order you like to read them, with news stories chosen to match your interests. Or maybe you prefer the television in the morning, in which case the TV program delivers the news in the way you prefer: sports first, then weather, then local, and finally national and foreign. Or perhaps the stories are delivered to you in an order derived from the number of friends or colleagues who have "liked" or "recommended" them. A ding from your watch lets you know precisely when you need to leave if you're going to make your 8:30 appointment—relying on a real-time feed from traffic monitoring systems like Waze or future systems relying on V2V and V2I, discussed in the previous chapter. You step into the garage where your driverless car is already on, with the inside temperature adjusted perfectly to the outside weather and your personal preference. It zips away while you continue reading the news or perhaps take the time to return some emails or make some calls. You didn't have to worry about turning off lights, turning the thermostat up or down, or checking whether you have enough food for dinner. As devices in your home were digitized, sensorized, and connected, those tasks were turned into data, and algorithms are now automating them on your behalf.

Your pantry and refrigerator know which products are running low and which ones need to be re-ordered—which they do without your involvement.

Work Day: While at work, you're alerted at various times of the day about your home's activities. These tasks are all likely taken care of without your involvement, but you are given the option to override the automation. It rained the night before so your sprinkler system won't turn on at its normal time today—unless you decide to override the computerized recommendation, a choice you'll have with several other assorted options today. A package arrives at the house (via drone?), and you sign for it remotely. Your entertainment system (see chapter 9) tells you the next season of your favorite show is ready for streaming, while your calendar automatically, dynamically, and continuously updates itself for your evening commute to ensure that you aren't late to your daughter's soccer match or your son's baseball game. You can decide what you'll have for dinner, but nothing comes to mind. Your home recommends a few options based on what's in the pantry and refrigerator as well as your dietary and health needs and fitness goals.

5:39 p.m.: You leave the office in your driverless car after a long, hard day. You don't relax quite yet because you still have a few more emails to send and a call to make. Your wearable body sensors know that you have heightened anxiety and alert your home (and your other family members). When you eventually walk through the door, the lights are set at a relaxing level, while your favorite music is already playing. If you set your dinner to cook in the morning before you left (in a connected crock-pot, no less), it is ready and waiting.

9:00 p.m.: After you've watched highlights from the evening baseball or football game, you get an alert that your body is more tired than you realize and that going to sleep now would align with your body's needs, based upon your calendar of events for the next day. You don't fight it and crawl into bed.

You'll notice that throughout this scenario I'm intentionally avoiding predicting the details of specific "connected" objects and their functions. But with this sketch you can begin to understand how the home of the future will function: decisions driven by data. In the mind of science fiction writers and popular imagination a generation or two ago, these tasks were performed by androids. While I have no doubt that we'll see human-form robots in some way in our digital future, I'm less certain that they'll be doing stuff for us in the home. We don't need them. Through the Internet of Things and connected devices, the home will look and feel much like homes do today, but it will function as an extension of ourselves. Robots, as we imagine them, don't have much of a place in this scheme. That said, will your refrigerator be a robot? Is your driver-less car a robot? Those are far more relevant questions. In the sense that these objects will be automated things capable of independent operation, even communication, then yes, they'll be robots. But we shouldn't expect that the home of the future will have a robotic maid, like Rosie from the Jetsons.

Rather, the home, through its connected objects, will be attuned to our individual preferences, of both mind and body. The Internet will be there, but in the background, like a power source—funneling the data from one object to the next in a continuous flow. We likely will have some sort of PC, but so many other objects will replace the need to use this PC for Internet activities as we do today: reading the news, paying bills, checking bank statements, indulging our interests, interacting on social media. These things will be compartmentalized, separated from one another, as opposed to mashed together the way they are today whenever we go "online."

This compartmentalization is already occurring. We will see it developed to its fullest potential in the home of the future.

THE 3D REVOLUTION

One aspect of the new connected home that I deliberately left off our daily scheme is the 3D printer, because it's a topic that requires its own discussion. In fact we won't exhaust the topic of 3D printing in this chapter. Its ramifications will touch nearly every corner of our lives in the digital future.

Still in its infancy, the technology behind 3D printing has grown remarkably in the last few years. A $2.5 billion industry in 2013, it's expected to reach $3.8 billion by the end of 2014. In five years, according to some estimates, the 3D printing market will grow by 500 percent to $16.2 billion by 2018.[4]

That said, we're still decades away from each home having a 3D printer. But one day 3D printers could be as ubiquitous as paper printers—or even toasters—are today. The advent and wide use of 3D printers will cause a revolution in domestic life. Although the hype surrounding them is quite extravagant, 3D printing's possibilities really are practically limitless. In your own home you will be able to "print" nearly anything you want—given certain limitations, including cost of materials and the size of the object. At its most basic level, 3D printing will allow people to print out common household objects—lamps, chairs, tables, desks, and so forth. The first wave of printable objects will be small and simple, of course. But as the technology progresses so too will the complexity of printable objects—to the point where you'll be able to print electronics, hardware, even food. (The possibilities get even more remarkable when it comes to healthcare.)

What this will mean in practice is that "shopping" could be a thoroughly in-home experience. When you shop online for some things you won't be buying physical goods any more—you'll be buying 3D blueprints, specifications that you download from the manufacturer to your 3D printer. Filled with the right materials, the 3D printer makes the product for you. This doesn't mean you won't ever have to purchase physical goods again. But the technology makes possible a future decidedly different from our present.

When each home can print its own products, then each home becomes in effect its own factory. Once more, we're back to individual craftsmanship, when families made things for their own use. But is this the right image? The problem with pre-Industrial Revolution individual craftsmanship was the quality; it varied from item to item even for the same craftsman. A great advantage of 3D printers is medieval craftsmanship with Industrial Revolution consistent quality. 3D printers also put customization on an industrial level. Indeed, you will be able to customize the specs you download to fit your needs perfectly.

With 3D printing, you'll be able to size the model you like to fit your use. You'll be able to customize the color, the fabric, and maybe even the material. The items in your home, in essence, will be custom made for you—but without the price of bespoke. When the customization is really just data, anyone can buy it.

This is the theme of the home in the digital future: a home that caters to your needs and wants, serves as an extension of yourself, and removes the many frictions and obstacles that keep us from living our lives to the fullest.

CHAPTER 9

The Mass Customization of Entertainment

"The history of the Internet is, in part, a series of opportunities missed: the major record labels let Apple take over the digital-music business; Blockbuster refused to buy Netflix for a mere fifty million dollars; Excite turned down the chance to acquire Google for less than a million dollars."

—James Surowiecki

On June 3, 2009, I found myself sitting before the Federal Communications Commission about to give testimony on the status of the digital television transition. Readers may recall in 2005 Congress decreed that all free over-the-air television analog broadcast signals were to transition to digital by February 17, 2009. In late January Congress extended the deadline to June 12 over concerns that there were millions of Americans who were unprepared for the transition. Four months later the same concerns were being raised again, which is why I was about to testify to the FCC. I delivered CEA's message, which was pretty straightforward: relax, everything's going to be fine.

OK, I didn't use those words exactly, but pretty close. After explaining why reports—they were really rumors—of converter box shortages were unfounded, I concluded, "In short, CEA remains optimistic that the DTV Transition will go down in history as one of the most successful public-private partnerships in our nation's history. While no one can guarantee that 100% of the nation's viewers will have taken the necessary steps to prepare for June 12th, we can say that the vast majority of the nation's television viewers are prepared for the transition. All of the DTV transition stakeholders should be immensely proud of the work we have done together over the past several years."

We were right. Save for a few isolated pockets across the country, Americans handled the digital transition exactly the way those of us who had been pushing for it had predicted: they didn't notice a thing. One reason? Cheaper digital television sets. As I told the commissioners, digital television sales were up 32 percent year-to-date even in the midst of a deep recession. In fact, more than 112 million DTV sets had been sold in the United States, many to exclusively over-the-air households. While there were indeed millions of Americans who obtained the federal coupons to purchase a digital converter, Nielsen had reported that it was just 2.7 percent of U.S. households.

In any case, the trajectory for television in U.S. households was clear: the new standard was digital. It was only a matter of time before all U.S. households were digital television households, particularly since digital TV prices continued to plummet. In other words, consumers weren't the ones who had to be convinced of the benefits of digital TV; it was the broadcasters, who had battled the transition at every turn. Here was a case of an entrenched, legacy industry that refused to modernize *even when that's what consumers wanted.*

As CEA President and CEO Gary Shapiro said at the time of the transition, "I think broadcasters blew it in that HDTV was their one opportunity to get ahead of cable and satellite in the sense that it was cheaper for them to go to HDTV because they could just send out [an HD] broadcast signal. They just have to invest in the towers. It could have been their competitive advantage. With cable, everything they sell has to be in HDTV. And broadcasters did not push the concept of free over-the-air television and their market share has gone down still to this day dramatically. And along came not only cable and satellite but now the Internet, and soon mobile devices."[1]

To be fair to the broadcasters, they did fund a lot of testing and research of digital television signals. But the point stands nonetheless: the default position of most entrenched, legacy industries is to fight change. For a variety of reasons, this tendency is deeply rooted in the entertainment industry, not just among broadcasters, but also among content creators—the movie and music studios and publishing houses. We're going to get to those reasons in a moment.

Aside from a handful of low-power stations serving small and remote markets, by and large nearly all television broadcasting is now digital. June 12, 2009, nine days after my FCC testimony, marked the conclusion of a project that had started nearly thirty years earlier, when a U.S. consortium of private and public interests developed the best digital television standard in the world. As Shapiro, who had been involved with the HDTV effort from the very beginning, said, the DTV transition was "the equivalent of putting a man on the moon."[2]

Once developed, however, we had to get it into use. From the consumer standpoint, that didn't prove to be as much of a problem as some had imagined. Moore's Law, matched with the superior technology, all but guaranteed that analog television was dead.

Congress didn't kill it with the mandated deadline. Consumers killed it with their dollars. It was, to use a phrase one should not use when discussing history, inevitable.

But it was also very messy, and it didn't have to be. Yet we shouldn't expect otherwise, not when so many interests are on the side of the old guard. And the messiness continues today. Digital is moving through the entertainment industry like a bulldozer, leveling structures that have been in existence for decades. Nothing is going to stop the bulldozer, but that doesn't mean something won't try to get in its way.

FROM SCARCITY TO ABUNDANCE

Since the days of the ancient poets and playwrights, one characteristic has defined entertainment for mankind: scarcity. In ancient times, the cost of paper and the written word precluded an abundance of poets and writers. Only the best were published—and only the best have come down to us thousands of years later. So too with playwrights. There were only so many festivals and theater productions each year at which playwrights could showcase their skills. Only the best competed against one another in ancient Athens, and only the best created works survive to this day.

It has been no different in modern times. Of the thousands of books that are submitted to publishers every year, only a fraction ever see print. Of the thousands of screenplays that get written, only a fraction become movies. There is a limited amount of shelf space in the bookstores and theaters in the world. Only the best make it there; the rest are never seen.

In an analog world we attempted to overcome these constraints by increasing the volume of whatever the scarce resource was. So we went from single-screen theaters to dual-screen theaters to multi-plexes with ten or more screens. In publishing, the Internet eliminated some of the scarcity of the bookshelf, but as long as publishers were still using printers and paper and ink, there remained a scarcity of resources. In either case, the scarcity was perhaps improved, but it remained a scarcity.

In an analog world, the limiting factors of scarcity dictate the choices individuals make. Consumers only see and read what has already been decided by an elite class of publishers and producers to be the best (or at least the most profitable). These elite gatekeepers determined what was on our shelves, on our television screens, and in our theaters. Consumers did dictate which of those works that made it past the gatekeepers would be successful, but they were starting from a selection that had already been greatly winnowed by others.

In his seminal 2004 *Wired* article "The Long Tail,"[3] which he expanded into a book, Chris Anderson called this winnowing process "the tyranny of lowest-common-denominator fare," by which he meant that economics—not quality—determined what we know as popular culture. Scarcity dictated what theater and bookstore owners would stock. If a product couldn't pay its shelf-space "rent," it was discontinued. Of course this doesn't mean that the product in question wasn't good or popular with some—it just wasn't popular *enough*.

The same is true for analog radio and television. There are only so many stations, and each can broadcast only a set number of hours of programming. "[T]he tyranny of lowest-common-denominator fare" demands that in such a world only those programs that attract the most ears or eyeballs will be broadcast.

Of course the problem is, as Anderson wrote, that "everyone's taste departs from the mainstream somewhere." Whatever movie is the biggest summer hit, whatever book is No. 1 on the *New York Times* best-seller list, or whatever television show gets the best Nielsen ratings week after week. Everyone loves something that wasn't a commercial success. But in an analog world, where scarcity dictates what is available for sale, you can't read, watch, or listen to the non-hit.

Digital removes scarcity from the equation. With no physical limitations, digital libraries can carry anything and not hurt their revenue—in fact, sales increase. In an analog world, stocking a non-hit CD means that you consequently stock one less hit CD—you're losing money by catering to fewer customers. But in the digital world, your shelf space "rent" essentially drops to close to zero. You can carry both, attracting both customers, because your shelf space is limitless.

"A hit and a miss are on equal economic footing, both just entries in a database called up on demand, both equally worthy of being carried," wrote Anderson more than ten years ago. "Suddenly, popularity no longer has a monopoly on profitability."

If you chart the sales curve of any digital library—iTunes or Netflix, for example—you'll find that the "hits" still produce most of the sales. But while the sales curve drops down quickly, it never goes completely to zero, as there are thousands of consumers who want the misses. While not as widely popular as the "hits," the misses aren't misses in the absolute sense because they are "hits" to some. This is what Anderson called the "long tail," where consumers can find anything they want in a digital world. "Unlimited selection is revealing truths about what consumers want and how they want to get it in service after service," according to Anderson.[4]

There is a long tail in all things, and digital makes that long tail more accessible to the masses. When I was a kid and listened to certain niches within punk rock, I had to order online from distant labels or visit "underground" stores. There was no sampling of long-tail content in the analog world because distribution was often not close. I bought plenty of albums I didn't like. But I had no choice. When I wanted to discover new bands I had to take risks. I relied on the labels or other influencers to put out something in my genre, but many times had to just buy whatever looked interesting. The greatest inhibitor of reaching deep down the "long tail" of media entertainment wasn't a lack of interest from other potential buyers. There were markets, but interested buyers were geographically separated, so there was no critical mass of buyers who could sustain niche markets in a physical world. In the digital world, distribution is always at the end of a connection, and sampling can easily be had.

Ten years after Anderson wrote the *Wired* article, his thesis has been proven correct—to the delight of audiences and the dread of the incumbent content creators and broadcasters. For instance, domestic movie ticket sales, an example of the old analog model, peaked in 2002 at around 1.6 billion tickets. Ticket sales in 2013 were 1.3 billion tickets—a sizeable number to be sure, but we are talking about 300 million fewer moviegoers.[5] Thanks to the Internet (and really, the digitization of content), more content is being created, developing more viable markets today than ever before.

But the Internet has also led to a raft of disruption in the physical retail environment. Hollywood Video was sold to Movie Gallery in 2005 and ceased operating in 2010 when Movie Gallery closed and liquidated all of its stores, which at the peak had included more than 4,700 stores, under chapter 7 bankruptcy. Blockbuster Video was sold to Dish Network in 2011 and as of January 2014 had closed all of its company stores, which had numbered nine thousand stores at the peak in 2004. Bookstores and record stores are in equally bad straits. Record stores are all but obsolete—aside from the occasional vintage record store. The corner bookstores, once a universal feature of Americana, has experienced a decade of decay. Crown Books was liquidated in 2001. B. Dalton books, with nearly eight hundred storefronts at the peak, ceased operations in 2010. Borders Books filed for bankruptcy protection in 2011 and sold or closed its assets including over five hundred stores. The last remaining big-box book retailer, Barnes & Noble, announced in June 2014 that it was spinning its e-reader Nook division into a separate publicly traded company. Reacting to the split, one financial analyst told Bloomberg News, "the company has not had an easy time over the last couple years."[6] An understatement for the entire category.

It's not just retail stores, either. It's happening with *all* physical entertainment. In June 2014, PricewaterhouseCoopers (PwC) released a study showing that by 2016, electronic home video revenue will exceed that of physical home video (DVDs)—that is, streaming and video on demand revenue will exceed that from DVD and Blu-Ray discs. The report estimated that DVD revenue will fall more than 28 percent, from $12.2 billion in 2013 to $8.7 billion in 2018. According to the study, in 2018 subscription video-on-demand services and cable on-demand offerings will be the dominant contributors to total filmed entertainment revenue, including the box office. In five years, electronic home video revenue will double from $8.5 billion in 2014 to $17 billion by 2018.[7]

In August 2014, Netflix CEO Reed Hastings announced on Facebook (where else?), "last quarter we passed HBO in subscriber revenue ($1.146B vs $1.141B). They still kick our ass in profits and Emmys, but we are making progress. HBO rocks, and we are honored to be in the same league." HBO, of course, is the dominant subscriber channel on cable—and no defender of the old guard. Its

TV series have been some of the best in the industry, and it is embracing the new digital "shelf-less" space with its streaming service "HBO GO." In late 2014, HBO announced they would finally offer an over-the-top streaming service outside of the traditional cable subscription model. HBO CEO Richard Plepler commented at the time, "In 2015, we will launch a stand-alone, over-the-top, HBO service in the United States.... All in, there are 80 million homes (of a possible 115 million) that do not have HBO and we will use all means at our disposal to go after them."

Much of this is old news. Physical entertainment—even digital physical entertainment including DVDs and CDs—will succumb eventually to the digital bulldozer, just as Anderson predicted. Services restricted by 24-hours-programming limitations that inhibit them from finding the addressable markets will have to change their ways or fail. Unable to compete against a business model whose "shelf-space costs" are zero, not to mention unable to carry the variety consumers have come to expect, physical distributors and their products find themselves drowning in a world of digital, streaming abundance.

THE WHOLE WORLD A STAGE

We're going to return to the travails of the content industry, but before we do, we need to look at another consequence digital will have on entertainment—something that not even Chris Anderson truly appreciated, at least at the time he wrote his groundbreaking article.

In 2006, entrepreneur and author Andrew Keen wrote an article for the *Weekly Standard* titled "Web 2.0." In it he wrote that the new Internet, or Web 2.0., "worships the creative amateur: the self-taught filmmaker, the dorm-room musician, the unpublished writer. It suggests that everyone—even the most poorly educated and inarticulate amongst us—can and should use digital media to express and realize themselves. Web 2.0 'empowers' our creativity, it 'democratizes' media, it 'levels the playing field' between experts and amateurs. The enemy of Web 2.0 is 'elitist' traditional media."[8]

In short, the Internet, powered by new digital tools such as streaming video, self-publishing, and musical software, gives everyone a stage, their own personal "American Idol," but without the need to pass through any sort of audition period. There you are, with an opportunity to be heard, seen, or read by millions, at virtually no cost other than the time spent producing your work.

Artists who otherwise would never have had a chance can make it big because of the opportunities afforded to them via digital content. Two notable examples come to mind. Before the pop singer and teen heartthrob Justin Bieber made it big, he was just a Canadian kid with a decent voice whose friend posted some of his singing-competition performances on YouTube. In 2008, a year later, a record producer just happened to click on one of Bieber's performance and, impressed, decided to track him down. The rest is history.

Then there's E. L. James, an amateur writer who worked as a studio manager's assistant in her native England. A fan of the vampire-based *Twilight* series of books, James began writing for the FanFiction website, using characters from the series. The sexual nature of her fan fiction angered the site's readers, so James began publishing her stories on her own sites. Her underground popularity caught the attention of a publisher, who worked with James to remove all the *Twilight* influences, renamed the first book *Fifty Shades of Grey*, and published it in 2011. Two more sequels followed.

To date, James has sold more than 100 million copies of the series, which is now also being made into a movie trilogy.

Of course, most amateurs don't get anywhere near such success, and most professionals don't either! But it's not so much the chance at phenomenal success that lures the "creative amateur" to the DIY model online; it's the chance for any success at all. The beauty of Anderson's "long tail" is that you don't need to be a mega hit anymore to be a professional. Talent alone can get you there—even if you still need a little bit of luck.

Of course, there's another perspective. *Are* Justin Bieber and E. L. James any better than the hits of the analog era? They seem to be producing the same kind of mediocre pop cultural fare that caters to the "lowest-common denominator," in Anderson's words. What about the rest of the creative amateurs? What's the chance that a Mozart, a Hemingway, or a Sinatra is really producing ingenious works in obscurity? Say what you will about the elite gate-keepers, they at least allowed some great talent through their gates.

And maybe the possibility exists now, with so many voices and works now vying for attention on the digital shelf, that the really good ones get lost. Why read Shakespeare when you can indulge your erotic fantasies in E. L. James's stories? Why listen to Bach when Bieber is so much less demanding? For the amateur artist, the elite gatekeepers were an obstacle; but some will argue that for the rest of us, they provided an extremely useful service by ensuring we had a chance to see, listen to, and read not the "lowest-common denominator," but the very good stuff.

This idea that by democratizing content we're losing the very scrutiny that allowed the truly great works to be created in the first place is the reason Keen in the end lamented what "Web 2.0" would wreak on society:

Is this a bad thing? The purpose of our media and culture industries—beyond the obvious need to make money and entertain people—is to discover, nurture, and reward elite talent. Our traditional mainstream media has done this with great success over the last century. Consider Alfred Hitchcock's masterpiece, *Vertigo* and a couple of other brilliantly talented works of the same name Vertigo: the 1999 book called *Vertigo*, by Anglo-German writer W. G. Sebald, and the 2004 song "Vertigo," by Irish rock star Bono. Hitchcock could never have made his expensive, complex movies outside the Hollywood studio system. Bono would never have become Bono without the music industry's super-heavyweight marketing muscle. And W. G. Sebald, the most obscure of this trinity of talent, would have remained an unknown university professor had a high-end publishing house not had the good taste to discover and distribute his work. Elite artists and an elite media industry are symbiotic. If you democratize media, then you end up democratizing talent. The unintended consequence of all this democratization, to misquote Web 2.0 apologist Thomas Friedman, is cultural "flattening." No more Hitchcocks, Bonos, or Sebalds.

Given that Keen wrote this lament almost a decade ago, we should now be living in the world he dreaded. So are the entertainment offerings available today worse than when the elite gatekeepers managed the flow of content? Another question: With so much more content on offer, are the classics of literature, music, and film now buried beneath a mountain of mediocrity?

The answers to those questions may depend on your own tastes and attitudes, but there is at least one medium that has flourished in the digital world—precisely because the creators behind the content were able to grasp the opportunities that digital provided. That they even saw them as opportunities is in itself a marvel, since so many of their fellow content creators have looked at the same set of facts and concluded that the digital wave must be opposed at all costs.

I'm speaking about television.

THE GOLDEN AGE OF TELEVISION?

When the final episode of *M.A.S.H.* aired on February 28, 1983, 105.9 million viewers, or 60.2 percent of U.S. households, tuned in. It stands as the single biggest television broadcast ever, surpassed only by Super Bowl XLIV in 2010. On May 20, 1993, the final episode of *Cheers* aired, reaching 80.4 million viewers. On May 14, 1998, the finale of *Seinfeld* was seen by 76.3 million viewers. On September 29, 2013, the finale episode of *Breaking Bad* aired, attracting 10.3 million viewers. It was the third biggest audience for a cable finale, behind *The Sopranos* in 2007 (11.9 million) and *Sex and the City* in 2004 (10.6 million).[9]

What's going on here? Setting aside the obvious differences between broadcast and cable television, you can see a gradual decline in viewership for the last episode of the most popular shows. Was *M.A.S.H.* the greatest television show of all time? According to viewership, yes. Even according to the critics, it was a great show. But does a show's audience really determine its greatness?

In 1983, U.S. viewers only had a handful of other television options that night. How is it fair to say that *M.A.S.H.* is better than *Breaking Bad* when the latter had to compete against *hundreds* of other options? Put simply: from *M.A.S.H.* to *Breaking Bad* the television world exploded in a cornucopia of alternatives. It's not disrespectful to *M.A.S.H.* to point out that had there been a channel devoted entirely to sports (say, like ESPN), then the viewership for the finale that night would have been substantially less than it was.

The phenomenon that we've seen play out in television was aptly described by Eric Schmidt and Jared Cohen in *The New Digital Age: Reshaping the Future of People, Nations, and Businesses.* As they predict, "individual agency over entertainment and information channels will be greater than ever, as content producers shift from balkanized protectiveness to more unified and open models, since a different business model will be necessary in order to keep the audience."[10]

In other words, the real development in television since the last episode of *M.A.S.H.* is not only that there are more channels; it's that each of those channels caters to a specific audience. In 1983, you watched *M.A.S.H.* even if you weren't particularly fond of the comedy, the characters, or the story. Without any channel catering to your preferences, there was little other choice. Besides, *M.A.S.H.* was good enough—an approach that defined television programming for decades: good enough for the largest number of people.

Had *Breaking Bad* been on in 1983, there's a chance it would have been a popular show in that it attracted tens of millions of viewers each episode, but I doubt it. A high school science teacher who becomes a murderous drug overlord isn't exactly good family entertainment. Which is why it's all but certain that *Breaking Bad* could never have been made in 1983—but *M.A.S.H.* could.

What we're seeing with digital is that television creators can produce a show that appeals to only 10 million viewers and still be considered a phenomenal hit. With the exception of the Super Bowl, we'll likely never see a *M.A.S.H.*-sized audience again. Yet that doesn't mean that what's on television is mediocre fare.

In fact, what's on television and cable networks these days is quite exceptional fare. From *The Sopranos* to *Mad Men* to *Breaking Bad*, television is experiencing something of a Golden Age in quality of content. Producers, who for decades had to create shows that were "good enough" for most people, can now create shows that are "amazing" for a small number of people. Audience segmentation has unleashed the creative genius of the television format. Unbound by the confines of mass-market appeal, content creators can focus on serving a specific market.

Of course the Golden Age doesn't stop with cable programming. Indeed, it's just the beginning. The first season of *Breaking Bad* limped to the end with fair-to-poor ratings. But through social networks and the possibilities inherent in on-demand streaming video, the show caught fire before its second season. Millions of Americans who had missed the first season decided to give the show a try—something that was not possible just a few years ago. "Binge watching" is the pop culture term that marked *Breaking Bad*'s meteoric rise from season to season.

From a storyteller's perspective, this development is a godsend. Whereas writing a show for a broadcast station means capturing an audience within the first two or three episodes, unlinking television from a strict schedule allows writers to develop stories in a more traditional way: through character, through drama, through conflict. Shows today have much more freedom to develop as the story dictates, not as the near-impossible demands of the old analog linear programming dictate.

That's one reason that television shows today are so much better: because the creators behind them are able to actually tell the story they want to tell without having to worry about all of the limitations that the analog era placed on the mass production of content.

Taking this phenomenon one step further, Netflix—a medium entirely unbound from scheduled programming—began producing its own shows. The first, *House of Cards*—an American take on a British original with a deeply cynical view of politics—has become a mega-hit. Releasing an entire season all at once gives viewers the chance to watch when it fits their schedule, and frees a show like *House of Cards* from an episodic story-telling structure, which is a very limiting way for writers to work—it's the whole season that's the story.

Netflix, of course, had every reason to expect that it could produce a successful show. Why? Because it knows *exactly* what its audience is watching. Every time any show is streamed, Netflix can capture that data. It can tell which shows are popular with which audiences down to the individual viewer. Nothing of this sort was possible under the old analog broadcast model. Again, producers back then had to work with the "good enough" for the largest number of people. Netflix, however, can produce a show that's "great" for a small number of people and have itself a huge hit.

In short, we've seen the future of television and it is the Netflix model. The old linear analog broadcasting model of programmed shows set on weekly schedules was always a "good enough" approach. It forced viewers into artificial viewing habits. Rather than indulge in entertainment on their own schedules, people had to conform their schedules to match the television schedule. Streaming video reverses this entirely—people can watch when they want to watch. Again, the digital data has provided humans with a greater degree of personalization and choice, both for the viewers and the creators.

As the Netflix model becomes the dominant entertainment model, we can expect to see the quality of the shows increase. Viewer data, including preferences and demographics, will allow creators to target any audience with a show all but created for them. Indeed, there will come a time when viewers themselves will decide more of the direction of the show based on their reactions and responses to previous episodes and seasons. The creators, either wanting to steer the show toward what the viewers expect, or preferring instead to give them a surprise, begin to have a dialogue with their audience. This allows shows to be catered even more specifically to audience preferences and tastes.

Already we see greater viewer involvement impacting the progression of entertainment programming. "Reality shows" like *Dancing with the Stars* and *American Idol* allow viewers to vote who stays and who goes. Until recently this was done in a very analog way: using a 1-800 number that viewers had to proactively call in order to place their vote. Today, programs rely on social networking platforms like Facebook and Twitter where conversation can constitute a cast vote. In the future, plot-influencing data—whether deciding who stays and who goes on a reality show or more complex character development in a sitcom or movie—could be garnered from viewers more implicitly, by relying on personal health metrics such as heart rate and blood pressure and mood-tracking information captured through wearable devices. Not only would this provide real-time feedback, but since the data is linear it would also provide data in a continuum that could be mapped to every second of a program, providing feedback on every element of that particular program.

While television is pioneering this new format, publishing and music can certainly learn a thing or two. As Anderson's "long tail" shows, there is an audience for entertainment of almost any type or genre. The trick is figuring out how to engage that audience with top-quality material. Is the democratization of content a good or bad trend? In his "Web 2.0" article, Keen warns against the scenario "when ignorance meets egoism meets bad taste meets mob rule." Certainly mass opinion hasn't always been good and accurate and is unlikely to be so in the future.

But digital gives equal weight to all voices. So while the majority might be heard more frequently, it doesn't necessitate that the minority be completely closed off. Indeed, digital allows for a whole lot more mediocre and terrible material; but it also allows me, the individual consumer, to discover and experience the good material, the material that seemingly was made just for my tastes and preferences. Television would never have had its current renaissance without digital data. We shouldn't expect anything different with music and publishing.

STILL FRICTIONS IN THE MARKETPLACE

Despite the positive influences of digital technology on entertainment, there remain massive frictions in the marketplace. During a recent Thanksgiving Day weekend, I attempted to introduce my sons to the original three Indiana Jones films. After creating excitement for what (in my mind) are some of the best movies ever put down on celluloid, I went to download them. I mistakenly set up my three sons' expectations on the presumption that I could deliver. I started by going to Netflix, where the movies were only available on DVD. Then I tried Amazon Prime and iTunes. I went to IndianJones.com—which promotes only the newer (and far less notable) *Kingdom of the Crystal Skull*. I tried the site for Lucas Films to no avail. My willingness to pay for a legal download continued to grow as I futilely searched.

It seems that even in a world of digital abundance, there remains scarcity. It's not just a problem with the Indiana Jones films either. It's a bit surprising that even in this day and age one cannot watch any professional sports game from the comfort of one's own home. The technology is there. Even the willingness to pay for it is there.

The frictions in the marketplace for digital content are largely the result of what economists refer to as concentrated benefits and diffused costs. In other words, this means that a few parties benefit at the cost of everyone else. Sugar policy in the U.S. is a prime example of this. U.S. sugar prices are wildly inflated compared to world sugar prices, because the sugar industry's lobby has convinced Congress to maintain strict restrictions on importation. As this phenomenon applies to digital data, we can say that while digital makes all these great things possible, we're still at the mercy of the law of concentrated benefits and diffused costs.

We have to understand that physical media—be it books, CDs, DVDs, or tapes—can easily be shared. Each book or record can be used extensively for years and then simply given away to someone else. But digital assets can be tracked and controlled. Copyright holders naturally want to maintain control for as long as possible and extract as much rent as possible. And so they exercise that power ruthlessly, often to the dismay of consumers, who can't understand why something as simple as a baseball game is unreachable without subscribing to an entire cable service.

With physical media, control was virtually impossible—despite those FBI warnings that came with each VHS cassette. However, monitoring and intervention are entirely possible in the digital world. This is often the reason rights holders seek power over distribution mechanisms such as websites, cable providers, and file sharing sites. These properties can monitor how the content is shared and restrict access to anyone who hasn't first ponied up. Rights holders would do this directly themselves, if they could.

It is impossible for the rights holders to alter physical assets after the purchase transaction. But with digital assets this is entirely possible. Digital assets resting on devices that are connected can be updated or changed. They can even be entirely removed from your device. This was the case in 2009, when Amazon remotely deleted copies of George Orwell's *Animal Farm* from individual devices. The digital asset in question had been sold by a company that didn't have the rights to the asset. Users did receive a refund. While Amazon has officially changed its policy on how it will handle similar issues moving forward, it is still the case that Amazon can change any of your digital assets lying on connected devices—up to and including full deletion.

Naturally distributors are just as covetous of their ability to control content as the rights holders. Without control, they lose all. Which is why we see broadcasters, movie and music studios, and publishers jealously guard their entrenched placement in the old analog model of content creation and distribution, even in the face of overwhelming and unstoppable trends.

As I said earlier, it's hard to blame them for their obstruction. We can, however blame them for refusing to innovate themselves. The broadcasters, for example, had every reason to adopt digital television early, yet they refused out of a short-sighted regard for their current monopoly. This not only hurt them in the end, it also made what could have been an easy and relatively painless transition to digital messy and drawn out.

At the heart of this friction are U.S. copyright laws. As we have seen, Larry Lessig has pointed out that throughout most of history copyright played a minor role in how ordinary people engaged with their culture. With the limitations of physical media, and a vice grip on content creation, copyright holders were for the most part protected from large-scale abuse. The average person, who might "steal" a physical copy of a tape by dubbing it, otherwise had no real means to violate copyright laws. Copyright regulated a very small percentage of our interactions.

Digital changes everything. As Lessig explains, "Copyright now reaches across the spectrum of ways in which we engage in our culture."[11] From music to social media to cutting-and-pasting an online newspaper article, you're engaging with—and likely infringing on—copyright. The entire Internet is one large copyright playground, all because of the properties of digital, which is easily replicated and unwilling to accept boundaries or limitations. It wants to be shared, just as much as you want to share it. To quote Jessica Litman again, "Most of us can no longer spend even an hour without colliding with the copyright law."[12]

A LOOK AHEAD

The future of the entertainment industry, in whatever medium we're talking about, is one of a dizzying array of possibilities:

Creation: Digital creates a mechanism for a feedback loop between content creator and content producer. Historical content creation relied on small focus groups to determine how different scenes might play in front of larger audiences. For example, in places like Las Vegas and Austin there are testing centers for content because Las Vegas and Austin are considered "average" America. As I pointed out earlier, content was historically made for averages. Digital changed that, and digital is changing it further. Now content creators can see changes in preferences materialize in real time.

One obvious way creators do this today is through social networks. When a television show displays a hashtag at the bottom of its program, its primary purpose is marketing, to create a "trending" topic on Twitter and foster a community; but its secondary purpose is for the creators to mine what fans of the show think about the episode. Do they like this character? Do they like that story line? What do they think will happen next week? Creators use this real-time feedback to help guide the future of the show.

And it's a practice that will only expand as the two-way communication between creators and viewers grows. Whether it's through social networks or some other medium, creators are going to want to be as close to their viewers' thoughts as possible—not just to determine if what they've done is working but also, as we have seen, to determine the future direction of the show.

Amazon recently released pilots for a number of series under consideration for original content production. The company let its entire user base dictate its future production investments—much as Netflix is able to determine the tastes and demographics of its users and craft a show it knows they will like. The metrics and digital data available to aid creation are set to explode as we digitize greater swaths of our everyday lives.

In short, the wall separating creators from viewers is getting thinner and more transparent every year. There will come a day when the wall disappears entirely and viewers will be taking an active role in the direction of their favorite shows.

Discovery: Today Netflix offers recommendations of movies you might like based upon some basic demographic details it has about you combined with your viewing habits and the viewing habits of its millions of other users. This is a huge leap forward from where we were. Digital is almost always at the root of recommendation engines. But this is just the beginning. As we increasingly digitize our physical space, the number of inputs into these recommendation and discovery engines increases exponentially. Imagine—not too many years from now—that you have a few friends over and you want to watch something you'll all enjoy. In the analog world you flipped through whatever was being pushed out across a few channels. The advent of cable opened up the choice but didn't help with the decision. Then Netflix came along and began offering recommendations, but those recommendations are based on your viewing habits which might be—and likely are—different from everyone else's in the room.

But now imagine that the Netflix home screen taps into other data sources in addition to your viewing habits. Perhaps it can also see the viewing habits of your friends by identifying them through their phones (which all have the Netflix app). It can pull data from the camera built into the bezel of your television and know how many people are in the room. It can see if you are lying down or standing up. It can pull weather-related information from your Nest thermostat and know how cold it is inside and what the weather is like outside. It might also pull in very personal information about you and your friends using your fitness trackers, such as how active you've been today. It might use heart rate and blood pressure to determine if you are excited or depressed. Then pulling in all of these diverse digital streams of data, it might be able to make movie recommendations that fit the people in the room and conform to your environment. While you personally might not choose to watch a movie like, say, *Napoleon Dynamite* by yourself, given the right environment, with just the right group of friends, it might be the perfect fit. We stumbled over these decisions in the past. Serendipitous discovery will still happen in the future, but it can also be significantly augmented by digital data.

To get discovery just right will require a delicate balance of security, privacy, and trust. While any of these metrics by themselves might not be very valuable or sensitive, taken together they paint a picture of us that could very well be a better depiction than we ourselves could provide. Certainly they best anything we could have provided in a purely analog world. How much information is enough to improve the recommendation engine without imposing costs on our privacy? In addition to counting the number of individuals in a room, what if the camera in our bezel also identified the clothes we were wearing or the furniture in the room? (What it cost us. When we bought it.) What data improves the recommendation algorithm and what data doesn't provide sufficient incremental predictive value is a cost-benefit analysis like none we have ever seen. Where the line is drawn is one of the pivotal questions today. We explore privacy in more depth in a later chapter, but it clearly becomes paramount in nearly every discussion as we press towards our digital destiny.

Niche corners: Today, millions of viewers tune into Twitch to watch others play video games. Yes, you read that correctly. There is a site that thrives on letting people watch others play video games. And YouTube has become a key destination when people are looking for "how-to" videos. There is literally no topic too obscure or mundane to have a devoted following. Go ahead, pick an obscure skill and search YouTube for it. See how many hits you get in return. Pentak Silat—a martial art from certain Southeast Asian island nations—yields 140,000 hits. Even out-of-date skills get hits. "Conjugating Latin" yields 1,400 hits. "Adjusting the tracking on a VCR" yields 764 hits. "Reading neumes"—useful if you want to learn Gregorian chants—yields 78 hits. It would probably help to know Latin for this last one too, so check out the Latin videos you found in the above-mentioned search. This is all going to explode even further. The Internet, and particularly the new Internet, is a perfect vehicle for bringing together people from all over the globe, people who are otherwise strangers, to share in their one unifying passion. Indeed, the social networks of the future will coalesce around these communities, catering to niche audiences. We already see this today with online book, hobby, and gaming clubs. The ability to indulge one's interests and passions will be no farther than a click or voice prompt away. Whether universal sites like Facebook or Twitter will split apart on account of this process or not, they will almost certainly lose their dominance as people choose to customize their social circles even more intimately. So instead of having one Facebook feed for all of your friends, you will have a feed for each of your niche communities.

CHAPTER 10

Healthcare in a Digital Age

"Next-generation medicine will utilize more complex models of physiology, and more sensor data than a human MD could comprehend. Much of what physicians do (checkups, testing, diagnosis, prescription, behavior modification, etc.) can be done better by sensors, passive and active data collection, and analytics."

—Vinod Khosla

The hit television series *House* had a long-running gag. If you don't know the show, the formula was pretty straightforward: a patient was brought in with unusual symptoms and Dr. Gregory House, a brilliant if rather temperamental physician, and his colleagues try to diagnose the mysterious illness. Once the team hit a wall—which they almost always did—they would start throwing out wild suggestions in exasperation. One of the wild suggestions that would usually get blurted out was lupus, which is one of those conditions whose symptoms can be so generic that they can pretty much fit any disease. An annoyed House would then respond, "It's not lupus—it's never lupus."

I have a sneaking suspicion that the gag was really a joke on the rest of us patients. As a population, we seem to believe that the Internet has given all of us honorary medical degrees. All it takes for self-diagnosis is a set of symptoms, Google, and a WebMD page. We rifle through symptoms in a desperate search to find out what's wrong, only to hit upon the devastating conclusion that we have lupus. Frantic, we immediately visit our physician to tell him the terrible news that we've contracted lupus—only to be informed by our physician, usually after a very brief checkup, that "It's not lupus—it's never lupus."

This is the blessing and the curse of healthcare in the new digital era. The benefits are indeed incredible. We're a few clicks away from millions of articles and medical sites that provide information about every conceivable disease, ailment, and bug known to man. And who knows better what we're feeling than ourselves? With a few clicks of the keyboard and a simple slide of the mouse, we can instantly grant ourselves medical degrees via the Internet.

In 2013, Pew Research reported that 59 percent of Americans have looked online for health information in the past year, while 35 percent say they have used the Internet to "self-diagnose." Interestingly, 41 percent of the self-diagnosers had their condition confirmed by a physician—a number that suggests people aren't as bad at self-diagnoses as perhaps many doctors believe.[1] Still, six in ten Americans incorrectly self-diagnose themselves, and what happens when they do is not known or reported.

Despite the danger of being wrong, the desire to know immediately what's ailing us and the temptation to use the Internet to find out are often too irresistible. That our personal medical quest usually takes us to the doctor's office anyway somewhat limits the dangers of pretending we're doctors—but not always. A sick patient can convince himself he is just fine after reading something online. Or perhaps the sick person comes across a pseudo-medical site that claims to have a non-medical solution to the ailment.

In any case, the Internet has empowered patients with an unprecedented level of data and health information. But like so much of the current state of digital data, healthcare is in the middle of the chaos. As patients, we may know more—or think we know more—than in the past, but we're still unable to use that information effectively on our own behalf. In other words, if we decide that after checking WebMD we should visit the doctor (a case of using the Internet in the right way), our medical experience is ultimately largely unchanged. The data pushed us toward a particular action (going to the doctor) but the doctor's visit experience, where we get the real care, remains the same.

But it's this experience—receiving the actual care—that is rapidly changing. Thus far, digitized data has changed only one part of healthcare services. But we are approaching a moment when digital data will have a pronounced impact on the entire healthcare cycle: digital data, the patient, and the treatment will reinforce each other, not only in better treatment, but also in helping us change our behaviors to live healthier lives. In this way, digital data is set to influence how we approach health and wellness from a number of fronts. From the doctor's office to the hospital to the gym, across the entire healthcare spectrum, this new level of care will represent a revolution in society.

For this chapter, we're going to look at three places where digital data will have the most pronounced effects: patients, doctors, and hospitals. Understandably, given the vastness of the healthcare field, we won't be able to cover everything. But we will see how the digitization of data is transforming healthcare through customization and empowerment for patients, doctors, and hospitals.

DATA AND THE PATIENT

Do you track any of your health information, such as weight, diet, exercise, blood pressure, or sleeping patterns? Chances are you track at least one of these, and it's probably your weight—not only because it's the easiest to track but also because it's the most visible indicator of our relative health. According to Pew Research, 70 percent of Americans track at least one health indicator for themselves or a loved one. It seems like a good, high number, doesn't it? Except that Pew also found that of those who track at least one health indicator, 49 percent do so "in their heads."[2] That isn't as encouraging; one begins to suspect that people might have been answering in the way they thought they *should* answer.

Of course the problem with tracking your health indicators is that there really isn't an easy way to do so—"easy" meaning having someone else, or something else, do it for you. You have to step on the scale (a painful process for some of us), write down your diet every day, monitor your work-out routines, sling your arm through the diastolic machine at the pharmacy—all of this is rather cumbersome for the average person. Certainly it is prone to inconsistencies. For your tracking to be effective at all—to give the data any import— you not only have to write it all down, religiously, but you have to constantly analyze it and attempt to decipher what it all means. Are things improving? Are they getting worse? Where was I successful and where did I fail? This is tracking your personal health in an analog world, and that's where I expect we lose most people.

After all, it's far easier to forgo self-tracking and just make an annual physical with the doctor. If everything checks out fine, off we go, happy in the knowledge that whatever we're doing, it's good enough and we'll just keep doing it. If things aren't fine, we promise to change diet and work out more, or we simply voice a painless "I'll take better care of myself, doc, I promise." It's no wonder obesity and diabetes rates are through the roof.

The problem with this common approach of yearly check-ups is that episodic recording of your vital signs doesn't help very much. It only helps if one of those signs, on that particular day, warrants medical treatment or a change in behavior. Thus health problems begin years before we—or our doctors—can identify them.

Consistent tracking, recording, and assessing of our health indicators would greatly help in identifying problems much earlier. But there are only a few of us who are so diligent.

The data is there, whether we track it or not. Our bodies are warning us about oncoming conditions well before they occur. If we could capture that data painlessly, wouldn't we? If we could take that data to the doctor, who would be able to analyze an entire year's worth of blood pressure statistics rather than just one reading every 365 days, wouldn't we? Even better, if we could painlessly have a doctor keep an eye on those metrics daily, wouldn't we?

Of course. And we are starting to, because the technology is finally matching our needs: the friction that used to exist between data capture and data recording is trending toward nothing. It's not quite there yet—the technology that is available, mostly through apps, still requires a great deal of human involvement. But this is the hybrid period between the analog world we knew and the all-digital world we will soon find ourselves in. Things have already moved far beyond the days when the only way to track your health data was through expensive instruments and pen and paper.

For personal monitoring to be most effective, it needs to drive behavioral changes. With sensors, we've taken analog information and digitized it. In many cases, we are capturing information that is already there, but might not have been easily observable. It is data that was always there, but just wasn't being captured in a systematic way. The next key step is for the digital data to then inform and influence change in the physical analog world. In other words, we need a feedback loop. Something happens in the physical world, we capture information about it in the digital world and then apply algorithms to that information to influence what happens back in the physical world. This process repeats itself in a continuous flow of information and behavior change.

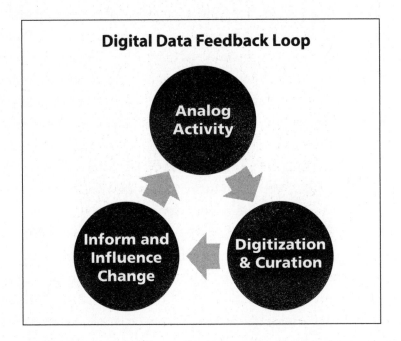

We should take a quick look at some of the more interesting products on today's market, because they give us insight into what to expect just a few years down the road.

iHealth manufactures an entire series of connected health trackers, from scales to fitness wearables to glucometers. Using an accompanying smartphone app, these devices not only capture the data, but then record and chart it. In an analog world, data capture was the relatively easy part—as long as you remembered to weigh yourself every day, you could just write down the reading in a notebook. But then all you had was page after page of numbers, useful perhaps in knowing the overarching trend of whether you were losing weight or not, but far from helpful in tracking the nuances of the trends in your weight. To do that you would then have had to chart the numbers manually or digitize them yourself by plugging them into an Excel spreadsheet. By this point, 99 percent of us have given up. It's too cumbersome to reasonably expect anyone to do it—certainly not consistently. And yet how valuable that data is! If charted, it could show us what times of the week, month, or year we lost or gained the most weight. That data could be lined up against other digitized streams of data like your personal calendar, which most people today keep digitally, to decipher correlations. Then we get to the real value of health tracking as I laid out above: influencing behavior. If we can say that our monthly buffalo wings night with the guys always seems to put on two pounds, then we know the behavior we need to change: stop eating the wings!

We can take this further by tying it back to our discussion of the home. Imagine a connected scale that is able to communicate to other relevant objects in your home, such as, say, your refrigerator. The scale relays your weight trend to the fridge, which, knowing your weight loss goals, warns you that the chocolate cake you're about to remove won't help you reach them. I envision my fridge yelling at me with the tone of a boot camp drill instructor. You too can have some fun thinking about the many snarky or cruel ways a refrigerator can help you keep on track; but the point is that this is how the digital data can influence your behavior—and influencing behavior is the key. By making sense of the data, digital technology helps you better understand what you need to change. But in the end—you have to do it.

Of course health tracking extends beyond our weight and vital signs. For instance, a variety of apps monitor sleep patterns using the sensors already on a smartphone. The early versions of these apps were a bit awkward because the user had to place the phone underneath the pillow or bedding—a rather annoying requirement. The company Beddit, however, now sells a separate ultra-thin sensor that users place under their bed sheet. With the accompanying app, the user can then track his or her sleep patterns, as well as heart rate and respiration. In the morning you can read your results, knowing quantitatively exactly how well you slept. And the accompanying smartphone app also suggests how to improve your sleeping patterns. Here again, we see that the key element is inducing behavior changes. Today you might share those results with a doctor, who would be able to analyze the results and recommend a personalized treatment plan to improve your sleep. Tomorrow, the smartphone app itself, using a myriad of other personalized data about things such as what we ate, when we ate it, and if we stayed up late binge-watching online content might be able to make even more robust recommendations about behavior change. Digital data is opening up an entirely new form of healthcare.

With a sleep app communicating with your iHealth app, you'd be able to discern whether your diet is affecting your sleep, whether your sleep is affecting your weight, or both. We're moving from simply tracking our health metrics in siloed systems to using the data in robust ways to drive behavior change.

The burgeoning wearable technology market is another source of health indicator tracking. By recording vital signs such as heart rate and blood pressure, a user can determine his or her level of stress, excitement, or anxiety. But the real value is when this data is shared between devices. So, for example, by blending your wearable health metrics together with your calendar, you can ascertain over time which meetings or other periods of the day give you the most stress. You might implicitly feel some stress during your evening commute, but blending a few digital data streams together will quantify the level of stress for you. Now add in digitized commute times and traffic patterns from services such as Waze, and we can start to build dynamic recommendation engines. You might get a recommendation to reschedule a meeting because the other person is also in a period of heightened stress. Or you might receive a recommendation to adjust your commute—say by leaving five minutes earlier or later to avoid heavy congestion—because the data suggests it won't delay your arrival by much but it should lower your stress metrics. Alternatively, your car's navigation system might take it upon itself to find a different route to avoid congestion and greater anxiety—continuously monitoring your calendar and stress levels to ensure the adjustments are effective.

There is of course the great promise of wearables for fitness activity levels and goals. Anyone who has tried a wearable health product like the Jawbone Up, the Misfit Shine, or the Fitbit recognizes the determination that overcomes you when you realize you are a few short steps away from achieving your activity goal for that day. Many can relate to getting up and walking around the house a few times just to hit some predetermined goal. Now imagine if we could apply that motivation to all aspects of healthcare. This is another example of simple data sets—in this case, the number of steps you've taken in a day—influencing behavior.

We are just now getting a glimpse of a healthcare future infused with data. I remember first seeing the Vitality Glowcap when the product was introduced by AT&T at the 2011 CES. While the

Glowcap may look like a regular pill bottle cap, it has embedded sensors and connectivity to the Internet through technology built into the cap. Digitizing the cap allows you to monitor when the bottle is opened and—more important—when the bottle is not opened. This information can be communicated to you, family members, friends, or others. Forget to take your medicine? The pill bottle can essentially call you. It can text you. And it can also alert a family member, friend, or medical professional who might urge you to take your medicine on schedule or otherwise check up on you. A push button on the bottle makes refills easier than ever as the required information is sent seamlessly to your local pharmacy.

By mixing multiple data streams, we can start to move away from simple correlation and look for causation. In an analog world, you see the end results of a multitude of factors—too numerous to analyze intuitively in your mind. But once you tie multiple data streams together, patterns start to emerge, and identifying what is causing those results is just a regression away. Digital data will allow us to take an active role in our health—not just by delivering us the right information, but by allowing us to see where our lifestyle choices influence our health. By tracking the number of steps I take in a day, I can see how my activity level is affected by, for example, going to bed thirty minutes later than normal the night before.

Another trend we're seeing in its earliest stages is remote medicine or telehealth. For instance, an app called First Opinion matches a user with a doctor whom he or she can text directly with questions. The service advertises that most questions are answered in under ten minutes—faster than you could even make a doctor's appointment by phone. You can text a photo of a rash, bug bite, or cut to the responding doctor for diagnosis or to determine if further medical treatment is necessary. Obviously, some questions will be unanswerable given the limitations of the medium, but rather than turning to Google, users can turn to a trained physician who will give them personalized medical advice. This is just one of many ways that extended reach of care is materializing.

A study conducted in Britain looked at the effects of telehealth technology on six thousand patients with chronic diseases. The study found that applying telehealth tools to this population reduced admissions to the emergency room by 20 percent and mortality by 45 percent.[3]

Looking down the road, we can extend this idea to virtual doctor appointments. A patient can transfer his or her relevant medical data (obtained via health tracking and monitoring devices) to the physician, who then determines whether a virtual appointment is necessary. Again, while there are some limitations, just a simple consultation with a doctor is sometimes all that is required.

People will likely always have to be near a hospital or trauma center of some kind; but virtual appointments will reduce the need to be within a reasonable driving distance of a family physician or a specialist. How often have you had to make an appointment with the doctor, which required taking time off work or taking the kids out of school, only to be told, essentially, "Come back if it doesn't get better"? These are what are known as "rejection visits"—you don't actually receive any medical care, except the assurance from a doctor that you aren't about to die. Imagine the resources and time that could be saved if "rejection visits" went digital, where virtual appointments decided whether an in-person appointment or even a hospital visit was necessary.

As I mentioned before, customization is one of the primary benefits of digitization and a key underlying theme of this book. Perhaps nowhere is customization more relevant than in healthcare, and yet healthcare is decidedly lacking personal customization. Like nearly every other manufactured product, medications are "one-size-fits-all." But the nature of medication is that not every person will have the same reaction. Drug companies and the Food and Drug Administration cannot develop drugs for each individual who's going to take them. Like the analog television broadcasters, they have to produce a drug that is safe and effective for the most people.

But what if they didn't have to operate under those restrictions? What if drug companies could produce a drug tailored specifically to your body, via your health data and genetic make-up? This is an example of personalized medicine—a developing field of medicine that uses a patient's genetic content to tailor treatments and medication to that individual. It's actually quite easy to know your own genetic sequence; the challenge confronting medicine is customizing drugs that work best with that sequence. Nevertheless, physicians could use a patient's genetic sequence to determine the best treatments currently available, as opposed to the trial-and-error process that guides most treatment decisions today.

An August 2014 study in the journal *Proceedings of the National Academy of Sciences* reported how researchers at MIT used RNA therapies in a mouse with lung cancer to slow and shrink tumor growth. "RNA therapies are very flexible and have a lot of potential, because you can design them to treat any type of disease by modifying gene expression very specifically," said an author of the study.[4] The first human genome was fully sequenced less than a decade ago. It cost $2.7 billion and took thirteen years. Since then prices have fallen below $5,000. Your personal genetic code will be a key piece of digital data, moving forward, and sequencing it is quickly moving within the financial reach of the developed world. Patients of the future will utilize this digital data together with physicians to craft personalized drug treatments for cancers and other illnesses: the patient's genetic code will be transferred to the drug manufacturer who will then make a drug tailored exclusively for that patient.

Finally, one of the great promises of digital health is suicide intervention. Today, the United States has nearly twice as many suicides as homicides, according to the Centers for Disease Control and Prevention. The August 2014 suicide of Robin Williams brought the danger of untreated depression to the forefront of the American public's attention once more, as many wondered how such a talented, successful actor could have ended his life so abruptly. The thing is, there were numerous signs that Williams was not in a good place mentally and, perhaps, physically—they were just hidden from almost everyone's view.

Digital data holds the promise of changing that.

Researchers at Northwestern University have tested a smartphone app called Mobilyze! to help with suicide prevention and the treatment of depression. Their approach uses sensors in your phone to monitor your movements—for example, whether you've left your house that day or moved around much. If the app senses that you've been idle too long, it will prompt you to call a friend or walk outside. Likewise, researchers at Dartmouth College developed an app called StudentLife, which uses the passive sensors embedded in a smartphone to monitor and measure stress levels. The software uses the accelerometer and GPS data to measure activity levels; the microphone detects if the individual is involved in conversations and social interactions; and the light sensor, the activity features from the accelerator, and the GPS and sound features from the microphone are used to monitor sleep. Taken together and monitored over time, these streams of data track changing stress levels.[5] But there are even more intimate ways of sensing a person's mood, particularly when it comes to depression and suicide risk. Social networks such as Facebook could one day determine your mental state just based on the number of posts you've put up or what kind of posts you put up. Wearable technology can already monitor heart rate and blood pressure, two indicators affected by more mental pressure. In all these ways, digital data can provide key insights and solutions for helping stem America's terrible rates of suicide.

Self-monitoring is going to continue to grow, because the healthcare system simply can't keep up with our diverse needs—especially as we age as a population. Helping to identify simple patterns will be a key function of this self-monitoring. Imagine someone falls at home and is injured but doesn't bother to visit a medical professional to get checked out appropriately. A wearable device with embedded sensors might "see" that the individual isn't walking as much or is sleeping longer and recognize these as signs of an injury and could then notify a medical professional or loved one.

DATA AND THE DOCTOR

Patients' ability to monitor and capture their health information is going to radically alter the physician's role. This has two consequences for health services moving forward. First, the shift to digital data monitoring means consumers won't be coming in for the most rudimentary tests. Doctor office physical exams, for instance, will likely decline. When patients can consistently track their own health indicators, there will be less need for doctors to do it once a year.

Second, patients will have access to services that once were the sole purview of the doctor. For instance, when X-rays were analog, they could be read only at the doctor's office. But digital X-rays can be read anywhere, and some facilities are already using health professionals in distant locales like India or Australia to read X-rays. In the future, the consumer can directly engage services scattered all over the world. Individuals need seek out local healthcare services only if the X-ray indicates additional procedures that can only be done locally. Economics teaches us that services will get both more efficient and less expensive when faced with global competition.

Of course this will impact doctors financially—at least in the short-term. Yet we shouldn't see this simply as another example of "shipping jobs overseas." Doctors, particularly primary care physicians, are already overworked. They spend countless hours seeing patients who don't require medical care and other countless hours in administrative work. By reducing their workload—in effect, handing that workload off to others—we allow the primary care physician to spend more time seeing patients and providing healthcare services.

Digital data will fundamentally impact the entire doctor-patient relationship. Today's doctor assessment largely relies on the same approach that has been used for hundreds of years. The doctor asks the patient questions in an attempt to pinpoint the problem by intersecting the answers with her own knowledge and experience. Of course this type of discovery is fraught with imperfections. The doctor can only go on what the patient says. The patient may withhold information either because he didn't think it important or out of a desire to keep something secret. But data never lies. Data will help doctors see things that you didn't think warranted telling the doctor and things that you wanted to keep hidden.

This suggests a fundamental shift in the traditional activities of the doctor. In the past, the doctor relied heavily on anecdotal verbalized data received from the patient. The doctor of the future might rely on reams of data captured over an extended period of time about a wide range of aspects on the patient. This development might require medical professionals to become data scientists first and caregivers second—with a key primary role of analyzing patient data sets, not only seeing patients. Of course seeing patients will still be a necessary function of the doctor of the future, but given the cost and resources now spent on routine check-ups (and especially "rejection visits"), data analyst is a role doctors will have to assume as digital makes the healthcare system more efficient.

Even though doctors may be seeing patients less, they will know a lot more about them—even details that the patient would rather keep hidden. But confronting uncomfortable truths about our actual health and lifestyle choices is part of effective healthcare. Digital data frees the doctor from trying to guess the patient's real behaviors.

In short, doctors will become more efficient. They will be able to take on more patients, as their per-patient hours shrink. Given the direction of healthcare, not just in the U.S. but across the world, this portends to be a very good thing. As a 2012 *Economist* article noted, "By 2030, 22% of people in the OECD club of rich countries will be 65 or older, nearly double the share in 1990. China will catch up just six years later."[6] With an aging population, chronic diseases become more widespread. In the U.S., for instance, basically half of all adults have a chronic condition such as diabetes or hypertension. Yet healthcare productivity has declined by 0.6 percent each year for the past two decades, as the *Economist* notes. And as *USA Today* reported in February 2014, the U.S. will have a shortage of 45,000 doctors by 2020.[7]

Given this state of affairs—an aging population and a rising doctor shortage—doctors of the future will need to be far more efficient. Digital data promises to provide at least a greater degree of efficiency, allowing doctors to spend less time with more patients.

DATA AND HOSPITALS

In no other field of healthcare are the promises of digital data as alluring—and as frustrating—than hospitals. To greatly simplify an extremely complex topic, hospitals produce and store tremendous amounts of data: data on patients, data on populations, data on diseases, data on asset utilization, and data on payments. In many ways, the whole notion of "Big Data" began with hospitals, which first started capturing and storing diagnostic and imaging data years ago. It was in hospitals, moreover, that digital data's promised improvement in treatment and cost savings were to be first realized through electronic medical records, and then through analytical tools that cut out redundancy and waste.

That was the idea, at least. Instead, for the most part, chaos has ensued. Spurred by federal mandates, hospitals have been transitioning to EMR and other data-first processes for several years. But with each healthcare organization moving in the digital direction at its own pace—and in many cases using different systems—seamlessness of data transfer and data use among and between hospitals has remained elusive. In many hospitals, huge amounts of data remain siloed in different departments, with no way to analyze that data or move it to where it needs to be: helping the clinician with treatments.

Beside the gargantuan task hospitals face, of not only making sense of the data they have captured and stored but also moving that data where it needs to be, some hospitals have resisted the data transition all together either because of cost constraints or a general resistance to change the systems that still work. The results have been healthcare costs that are just as high as or higher than before the transition to data began, and a patient population that has yet to fully appreciate the treatment benefits of big data.

The hospitals have had no shortage of help in this effort. Since 2010, according to a study by McKinsey & Company, more than two hundred companies have been created to develop the necessary tools and software to help hospitals manage their data transition.

Nevertheless, digital data promises to alleviate the pressures and trends that have long troubled hospitals. Looking at a few benefits of digitization for hospitals in more detail will help readers better understand the role that digital data will play in the future of healthcare:

Sustainable costs: By far the biggest problem for hospitals is how to control or reduce costs. Many hospitals and healthcare organizations are adopting risk-sharing plans, in which physicians are reimbursed not for the volume of patients they treat, but for patient outcomes or cost control. Thus it is in the physicians' best interest to reduce the number of treatments a patient receives while also producing a better outcome. This gives hospitals an incentive to capture and exchange patient data with payers. The changing reimbursement environment also entices hospitals to make treatment decisions based on the best possible scientific evidence. This demands access to big data sets, as the McKinsey report notes, "In many cases, aggregating individual data sets into big-data algorithms is the best source of evidence, as nuances in subpopulations (such as the presence of patients with gluten allergies) may be rare enough that individual smaller data sets do not provide enough evidence to determine that statistical differences are present."[8]

Improved care: A trend within larger hospital organizations has been the development of what are known as health-information exchanges (HIEs). This kind of organization is essentially a collection of hospitals that all use the same IT system to streamline data sharing and patient records. Within the HIE, therefore, a patient can be sent between specialists with little to no redundancy, as the patient's health record moves with him electronically. HIEs also expedite the sharing of knowledge about large patient groups, from which clinicians can pull relevant disease and treatment information to improve care. For example, if a patient comes to the hospital with complications resulting from diabetes, the clinician can access the diabetes database to find the best scenario, culled from hundreds of like cases, for this particular patient. HIEs also provide the working model for a nationwide healthcare system, containing thousands of hospitals, able to communicate and share data with each other.

Business intelligence (BI): Even small hospitals are large organizations, in that they employee hundreds if not thousands of administrators, physicians, nurses, and other staff. Managing such a large and complex organization requires a careful balance of resources and costs. Digital data helps hospitals make sense of their administrative needs through business intelligence, which is simply turning raw data into useful information for the purpose of business analysis. BI isn't exclusive to healthcare, but it's in healthcare that we can see the greatest application of BI tools. For instance, BI tools can allow hospital administrators to rate and evaluate physicians and nurses based on patient report cards given to the patient upon release. Understanding the patient experience, not just the quality of care, also helps hospitals understand their intake and release processes— how long it takes to admit patients and release them—as well as more general matters such as custodial and cafeteria care. Pulling this wealth of data together in order to find trends and make decisions is the whole idea behind business intelligence.

Predictive preventative medicine: Today we talk about preventative medicine and preventive healthcare as the things individuals do to minimize future medical treatment such as disease treatment and hospital stays. But the future of digital will enable a world of predictive preventative medicine, wherein personal data together with algorithms will help predict and prevent ailments (not just treatments) before they happen and thus enable us to undertake extreme preventative care.

Code blue is an emergency alert system used in hospitals to signify that a patient is going into cardiac arrest. It is a resource-intensive process that is both chaotic and intense. Unfortunately, the code blue patient has less than a 20 percent survival rate. But before reaching code blue level and going into cardiac arrest, research finds, patients already show signs of clinical deterioration. Sriram Somanchi, a Ph.D. candidate in Information Systems and Machine Learning at Carnegie Mellon, together with Samrachana Adhikari, Allen Lin, Elena Eneva, and Rayid Ghani recently published research showing that early prediction of cardiac arrest is possible using patient data such as demographic information, hospitalization history, vitals, and laboratory measurements. The researchers were able to predict code blue with around 65 percent accuracy and only 20 percent false positive rates at 4 hours before the event.[9]

The benefits of early detection are obvious. Hospital resources will be used more effectively and efficiently, and ultimately lives may be saved. The algorithms still need more refining—something that will naturally happen over time as more information is digitized.

A final word on the promise big data holds out for healthcare. We all know that the term is derived from the sheer size of the datasets. Yet it's something of a misnomer. Yes, the data is big. But it's really in the little decisions that come out of the big data (and powerful algorithms—don't forget the math!) that we find its awesome value and potential. It's the little decisions that big data makes possible—from brushing your teeth better to taking action when you are hours away from going into cardiac arrest—that promise to transform healthcare by empowering providers and customizing care for patients.

THE HEALTHCARE HURDLES

There is no shortage of challenges as we move from the healthcare world of today to the digital data healthcare world of tomorrow. Here are a few of the hurdles worthy of our attention.

Security: As with nearly every aspect of digital data, privacy and security are of great concern when it comes to healthcare. The concern around healthcare data is probably equal to none anywhere else outside of perhaps personal financial information. We regularly hear in the news about health record data breaches. This has been partly attributable to hospital administrators' unfamiliarity with electronic health records—something that will only improve with time—but it also has to do with the ease with which digital data wants to replicate. Hospitals are going to have to create and administer security systems that are just as advanced as the ones used by financial institutions. Patients are understandably sensitive when it comes to their personal health information, and this sensitivity carries over into concerns about participating with institutions that use digital health records. Hospitals and primary care physicians can go a long way toward easing patient concerns by implementing up-to-date security protections.

Disentangling wellness devices from medical devices: The ability to digitize information that has always been "there" but was not easily observable or recordable is a key theme of this book. This new capacity is creating entirely new industries, and the devices and applications that have arisen are impressive—especially in the fitness and medical space. When the smartphone first launched in 2006 most never envisioned it would one day be a hub for your fitness and wellness—from which you would check your blood pressure and monitor what you've eaten, how far you've walked, or your posture. These are all very real applications in common use today. And much more is coming. With simple wearable sensors you can measure breathing, pulse, or sleep state. The data being created is immense, and it will eventually inform many of the decisions that you will make, and that are made on your behalf. Across the entire spectrum of health-related metrics, we are seeing sensors integrated into a plethora of devices. This not only enables the digitization of data that was difficult to see, but also facilitates consumers' doing more than they've ever been able to do for themselves. In the future, consumers will be far from passive participants in their healthcare. The FDA should clearly distinguish fitness and wellness devices from medical devices. Medical devices will require lower false positive and false negative readings. But it is worth noting that the litigious U.S. legal system demands accountability even of inherently imperfect systems—ignoring the greater good that can come out of them. Tort reforms are needed to create a more appropriate balance between innovation and perfection.

Using anonymized data: The South Napa Earthquake that struck in August 2014 was the strongest to hit Northern California in some twenty-five years. Jawbone analyzed anonymized data from thousands of Jawbone UP users who were tracking their sleep patterns the night the earthquake struck at 3:20 a.m. In this way they were able to analyze how sleep patterns were impacted by the quake. This is just one example of thousands that show how aggregated digitized information can aid our understanding of the world in which we live. How to find the balance between siloed privacy and aggregated anonymized data analysis will be an important discussion in the coming years.

Insurance: With personal health information increasingly easy to capture and track, we are already seeing insurance companies using wearables as a way to incentivize consumers. Auto insurance companies have programs in place to lower your premiums if you are willing to share GPS information with them so they can confirm that you are a safe driver. Insurance companies want consumers who are healthy, so they naturally want to create programs for which the healthy will self-select. It's probably a stretch to think that they will ever use wearables as mandatory devices; rather they see them as tools to identify healthier—and thus cheaper—policy holders. Likewise, companies are able to lower their share of the insurance burden by having a healthy workforce.

PREPARING FOR THE REALITIES OF HEALTHCARE DEFINED BY DATA

Moving into a world where healthcare is governed by data presents a myriad of questions that pit the great benefits against the reality of potential costs. For example, what happens if my employer finds out I have a chronic health problem because of personal data provided to an insurance program or healthcare provider? It shouldn't impact me. But it might. This is the reality of a data-defined environment. Data-driven healthcare is a significant break from the way in which medicine has been practiced for hundreds of years. How do we prepare tomorrow's medical professionals for this reality? There has already been some initial integration into medical school coursework, but how do we continue to push this forward?

When everything can be measured, what should we measure? How do we make the great promises of a sensor-infused life consumable by the masses? How do we maintain consumer engagement when many metrics will rarely deviate from the individual's baseline? How do we make the data easy to share so that it can inform other aspects of our lives, while at the same time maintaining expected levels of both privacy and security? How do we balance the great potential of predictive medicine with the anxiety and costs associated with false positives and false negatives? Ultimately, how do we trust data, accuracy, and performance? CEA is working to help standardize sleep measurement definitions. It's a good start, on just one of many metrics we will self-monitor in the future.

As a society we're only just beginning to grapple with the consequences, positive and negative, of digitized healthcare. The realities of our digital destiny lie before us. But it's no overstatement to say that in such a highly charged area of our lives, where both the benefits and costs can be high, the coming debates will be fiery indeed.

CHAPTER 11

Of Politics, Data, and Digital Revolutions

*"This is not an Internet revolution. It would have happened any-
way. In the past, revolutions happened, too."*

—Wael Ghonim (2012), Internet activist and Head of Marketing,
Google Middle East and North Africa

A t 5:00 a.m. on January 18, 2012, more than 115,000 Web
sites went "dark" for twenty-four hours. Google, Wikipedia, Mozilla, Craigslist, Reddit, Flickr, Tumblr, Twitter, and
Wordpress were just some of the largest sites that participated in
the organized "blackout" to protest a pair of bills in Congress. Oh,
and the Consumer Electronics Association website went "dark" in
protest too.

But the effort wasn't limited to websites: 4.5 million people
signed Google's online petition opposing the bills; 8 million used
Wikipedia's site to look up their representatives' contact information; on Twitter, posts opposing the bill were sent at a rate of a

quarter of a million every hour, at one point constituting 1 percent of all tweets; and the congressional switchboard was fielding (or failing to field) two thousand calls a second. By the end of the "blackout," more than 162 million visitors had seen Wikipedia's "dark" site.[1]

Meanwhile, thousands gathered in protests across the country, including in New York, San Francisco, and Seattle. The inventor of the World Wide Web himself, Sir Tim Berners-Lee, speaking at an event in Orlando, Florida, called the SOPA and PIPA bills a "grave threat to the openness of the internet."[2]

In Congress, the focal point of this tsunami of digital activism, proponents of the bills began to waver. By the end of the day, ten senators and nearly twenty congressmen had renounced their support, including some who had been co-sponsors of the bills.[3]

A handful of the bills' stalwart supporters refused to believe their lying eyes. Some in Congress claimed that Wikipedia, Google, and a hundred thousand or so other opponents were spreading "misinformation."[4] Private-sector supporters of the bills called the entire episode a "dangerous gimmick," an "irresponsible response," and "an abuse of power."[5] But the facts were on the opponents' side.

And the online activists had succeeded. The bills, which had enjoyed widespread bipartisan support not a week earlier, were now dead; and they would remain that way.

It had all happened so fast that even the media struggled to make sense of it. A day after the blackout, a *New York Times* story attempting to reconstruct events quoted a New York University professor who explained simply that "a disorganized group of people online became a coordinated group of people taking action."[6]

Order from chaos.

THE MAKING OF AN ONLINE REVOLT

It is worth spending some time analyzing the SOPA/PIPA episode, not only because CEA played a role, but because it stands as a classic example of the way in which digital data has transformed politics. We'll discuss other examples in this chapter, but won't analyze each so minutely. An entire book could be written on politics and digital data, but this isn't that book. So as we look at other examples, we should keep in mind the framework of the SOPA-PIPA debate, which played out in the three distinct phases that characterize how political action will be disrupted as a result of the digital data's transformation of politics:

Phase 1: It begins with a relatively small niche community, which has a direct stake in the problem or issue or campaign—what economists refer to as concentrated benefits. This community represents the first network—sharing (that is, replicating) and analyzing relevant data, which is still mostly siloed within the community.

Phase 2: The community grows to include outside stakeholders—those whose interest in the problem, issue or campaign is less direct. The sharing and analyzing network grows—exponentially, as social networks do their thing.

Phase 3: The goals of the network are achieved—or not. The network then either disintegrates or recedes to its original size—relatively small and niche.

The two defeated bills, known in the House as the Stop Online Piracy Act (SOPA) and in the Senate as the PROTECT IP Act (PIPA), were the darlings of the content industry, notably Hollywood, music labels, and publishing houses. For years this coalition had lobbied Congress to enact tougher regulations and penalties against online piracy—a very serious issue. But in trying to combat the online theft that takes dollars away from content creators, Congress did a terrifically stupid thing: it all but stood aside as the lobbyists wrote the legislation.

The result was a proposed legal and regulatory framework that would have given the content creators near-complete authority to shut down sites they deemed party to piracy. It was as if no one in Congress had bothered to remember just how the Internet, particularly social networks and blogs, worked. To repeat the words of Larry Lessig, "in the digital world practically all uses of culture trigger copyright because all uses produce a copy."

It's not like this was unknown to the very protectors of copyright, the content creators. But rather than engage with the tech community—and those sites that depend on user-generated content, such as the gargantuan Wikipedia—they chose to go it alone—and to ram the bill through as fast as possible. Few in Congress, with notable exceptions such as Democratic Senator Ron Wyden of Oregon and Republican Representative Darrell Issa of California, put up any resistance to the momentum.

We at CEA were one of the first organizations to call out the innovation-stifling bills, as well as Congress' backdoor dealings with the content industry. We were, you might say, a part of Phase 1. For instance, in October 2011, several months before the online blackout, CEA hosted more than a dozen venture capitalists who were among the 130 entrepreneurs, founders, CEOs and executives who had written a letter to Congress opposing the legislation.

"As investors in technology companies, we agree with the goal of fostering a thriving digital content market online," the letter said. "Unfortunately, [PIPA and SOPA] will not only fail to achieve that goal, it will stifle investment in Internet services, throttle innovation, and hurt American competitiveness."[7]

CEA President and CEO Gary Shapiro was an early outspoken critic of the drafted legislation, writing in prominent outlets like *Forbes* and *The Hill*. Michael Petricone, CEA's Senior Vice President of Government and Regulatory Affairs, and his team worked the inner corridors, and inner circles, of Capitol Hill. CEA is proud of its early involvement in the anti-SOPA/PIPA effort, but CEA recognized the forces arrayed were powerful. As Shapiro wrote in *The Hill*, "The copyright lobby is pursuing this all-or-nothing approach because it believes it has key members of Congress on its side—and to a large extent it does."[8] A massive grassroots uprising was needed. We needed a spark to ignite those who would be the real targets of this legislation: the tech community, including but not limited to the millions who rely on community-sharing sites such as Reddit and Tumblr, which serve a more tech-savvy crowd than Facebook does.

And a grassroots uprising is exactly what happened.[9] Around the time CEA hosted the entrepreneurs, Reddit and Tumblr users began discussing the implications of the legislation. The conversation expanded to include Mozilla, TechDirt, Kickstarter, and a handful of other online communities. Over the weekend of November 12–13, a meeting was held at the Tumblr office in New York, where some eighty representatives from the tech community discussed a course of action.

On November 16, the day when a House Judiciary Committee held a hearing on SOPA, Tumblr featured a "Censored" banner on its sign-in page, which directed users to information about the bills. It was the first assault from the bills' opponents—and it was a massive one. At the time, Tumblr hosted 40 million blogs, and whenever one of those bloggers signed in, he saw the "Censored" banner.

At this point, in mid-November, the movement hit Phase 2— expanding beyond the niche communities to include the general public. On Reddit a board created to discuss the legislation was getting 2.5 million visitors a day. The wave was building, but before it crashed on Washington it hit the blog-hosting company, Go Daddy, the largest ICANN-accredited registrar of internet domains in the world and a supporter of the legislation. With Reddit and Tumblr users in high gear, a boycott against Go Daddy began in late December, while a "Google bomb" was also starting to remove Go Daddy from its No. 1 slot on Google searches. On Christmas Eve, Venture-Beat reported that Go Daddy had lost 37,000 domains in the previous two days.[10] So it came as little surprise that on December 29, Go Daddy unequivocally rescinded its support of SOPA/PIPA.

By the beginning of the new year, Wikipedia and Google were on board with the effort—which tells you a bit about how this movement evolved. It wasn't a top-down, industry-backed, heavily resourced campaign. It was grassroots in the truest sense of the word. On January 12, during the International CES in Las Vegas, Senator Ron Wyden, Representative Darrell Issa, and CEA President and CEO Gary Shapiro held a press conference to bring attention to SOPA/PIPA and the imminent danger to the Internet.

The venue couldn't have been better. CES attracts some 150,000 attendees and 5,000 members of the media. Although much of the niche tech community was already actively opposing SOPA/PIPA, the CES press conference further galvanized the larger tech community, as well as the broader public who would read the dozens of stories written about the press conference online. Here are just a handful of the national stories that the CES spotlight spawned:

- *Washington Post:* "At CES 2012, Proposed Anti-Piracy Legislation Is a Hot Topic."[11]

- *Forbes:* "Time Is Running Out for SOPA Opponents, Congressmen Warn at CES 2012."[12]
- Fox News: "CES 2012: Wyden, Issa Decry SOPA; Mildly Hopeful for Their OPEN Bill."[13]

CEA's efforts at CES put SOPA/PIPA onto the front pages, pushing Phase 2 to the general public, but what was needed was a massive "protest" that would send a clear signal to Congress. But for all the resources CEA could bring to bear—and defeating SOPA/PIPA was a top priority—the network of millions of users on sites like Reddit and Tumblr were needed to take up the cause.

Phase 3 began and ended on January 18, where this chapter commenced. We don't know with certainty what would have happened had Congress ignored the groundswell, but likely Phase 2 would have continued until a resolution was achieved.

In 2013, Harvard researchers conducted an in-depth look at the SOPA/PIPA debate and what they call the "networked public sphere." On the study's interactive website[14] users can see how the conversation spread from site to site, through more than nine thousand individual stories posted on media outlets and connected via links across a myriad of sites. What emerges in Harvard's time-lapsed interactive graphic is a phenomenon of an ever-expanding, more connected network that the researchers say "reflect[s] quite a different game from the one that has historically played out in the traditional, mass-mediated public sphere."

The accompanying study, which tracks the interactive graphic during its various phases of growth and connectivity, produces a number of key findings, among them that "Individuals play a much larger role than was feasible for all but a handful of major mainstream media in the past. A single post on Reddit, by one user, launched the Go Daddy boycott; this is the clearest such example in our narrative. But we also see individuals embedded in organizations that in the past would have been peripheral, who are now able to play prominent roles...."[15]

"A single post on Reddit." In analog days, that "single post" could have been a guy sounding off at his local coffee shop—the data is only received by those closest to him and it's a good bet none of it is recorded. But through the Internet, the data is immediately recorded, replicated, and received by those who either agree or are at least receptive. From there, there are no limits on how far the "single post" can travel. That is the effect "a single post" can have in the networked public sphere.

As the Harvard study points out, the SOPA/PIPA protest began with a somewhat unified coterie of web-savvy individuals and groups. Its growth was nurtured by a central, if not quite command-control, group of interested organizations. But once the momentum began—and the many "single posts" replicated into a larger focused force—the effort did indeed collect previously uninterested parties into a single bloc of effective action.

The Harvard study concludes:

> By the end of the 17 months under study, a diverse network of actors, for-profit and nonprofit, media and non-media, individuals and collectives, left, right, and politically agnostic, had come together.... This outcome represents the fruits of the online discourse and campaign by many voices and organizations, most of which are not traditional sources of power in shaping public policy in the United States...perhaps SOPA-PIPA follows William Gibson's 'the future is

already here—it's just not very evenly distributed.' Perhaps, just as was the case with free software that preceded wide-spread adoption of peer production, the geeks are five years ahead of a curve that everyone else will follow. If so, then SOPA-PIPA provides us with a richly detailed window into a more decentralized democratic future, where citizens can come together to overcome some of the best-funded, best-connected lobbies in Washington, D.C.[16]

I would add one edit to this otherwise sound conclusion. It's a mistake to believe that online activism always favors the weak against the "best-funded, best-connected." Sometimes it will; sometimes it won't. Nothing stops the traditional sources of political power from using the same digital tools. We should not limit our understanding of the transformation in political activity by seeing it in such black and white terms.

The SOPA/PIPA episode has rightly been seen as a watershed moment in online activism. However, what few have noticed is how the SOPA/PIPA effort was more than the "the web flex[ing] its muscle," as a *New York Times* article put it.[17] This attitude seems to confine the episode to some distinct and distant entity—"the "Web"—as if the ordeal didn't galvanize real people in real life. One wonders if the same headline writer who came up with "With Twitter, Blackouts and Demonstrations, Web Flexes Its Muscle," would have written, "With Downfall of Egyptian President Mubarak, Web Flexes Its Muscle"? Or, if in the aftermath of President Obama's reelection, the headlines would have read, "President Wins Second Term; Web Plays Big Role."

This opaque view of the Internet as somehow detached from real life—as if the Internet is still a curiosity—and imbued with a collective consciousness limits our understanding of just what happened on January 18, 2012, and in the Arab Spring, and in the 2012 election. The Web is a connector. As I wrote in the 2014 edition of CEA's 5 Technology Trends to Watch, "The Web as we know it was designed as a tool for managing, finding, and retrieving information. In 1990 vernacular this meant linking to documents stored on different computers, or, the hyperlinking of distributed documents. In blending the physical and digital worlds we essentially extend the original concept of hyperlinking to include physical objects."

In the application here, "physical objects" can be extended by natural logic to include people. The Web isn't some nebulous object. It is an inanimate structure made up of well-defined nodes. While the number of those nodes is growing rapidly as an increasing number of objects, including people, latch onto this inanimate structure, each of those nodes is still known and distinct. Although without the Internet there would not have been a SOPA/PIPA debate at all, we must understand that the Web is simply a device through which digital data is allowed to move without many restrictions. In this way we see, yet again, the circular pattern of people influencing data and data in turn influencing people—prodding each other forward in a continuous round. Data influences people to action—in healthcare, as we saw in the last chapter, in political activism, as we see here—moving inaction to action.

Digital platforms exacerbate and accelerate dissemination as data digitizes. Take, for instance, elections, which quite clearly are pre-Internet phenomena. Elections are no different today in function, but the presence of the Web, which releases data in much greater volume, has made them feel more intense. Because of the Internet, election "data" hits us on multiple fronts—when in the old analog days, that data would hit us on only a few fronts and often with much delay. We see it in our email, on the sites we visit, and in the media we consume. That's not much different in kind from when election data hit us in our physical mailboxes, on billboards and banners, and in our newspapers, radio, and televisions—it's just more extreme.

DIGITAL DATA'S EXTREME EFFECT ON POLITICS

If we look closely at the SOPA/PIPA episode, we can see that three key factors central to any political effort—motivation, communication, and coordination—undergo an intensification process as data is digitized. As we enter the digital era, their relative power in influencing political outcomes increases significantly—if not exponentially. In other words, what has always worked in politics will still work in politics, but the addition of digital data has increased the potency of certain ingredients because digital data can spread faster and target more accurately. The available metrics around digitized information allow for continuous adjustments.

The essence of politics is how to get a group of people (usually a majority, but not always) to work toward a common goal—whether that goal is to defeat or pass legislation, to win an election or to launch a rebellion or some other form of protest, civil or otherwise. Let's look at these key ingredients one at a time:

Motivation: Before stakeholders will commit themselves to a course of action, they need to know that what you're asking them to do is right. Not right in the moral sense, but right in the self-interested sense (which sometimes can be the same as the moral right). Sometimes it's right for a senator to vote a certain way because that's what will get her reelected. Sometimes it's right for an employee to stand a picket line because that's what he thinks will be right for his job. Sometimes it's right for an unemployed student in China to brave the tanks because that's what he thinks is right for his cohorts. In every case, something motivated the stakeholder to believe a particular course of action was right for him or her. But political motivation isn't just about convincing stakeholders to side with you. Motivation is also about taking action—going to the polls, calling your representative, even taking to the streets.

In the analog days, electoral motivation—what do people care about and why do they care about it?—was ascertained through telephone surveys and focus groups. Analyzing the responses, you shaped your message of motivation to appeal to the masses. For the tech community, the anti-SOPA/PIPA motivation was to show how the bills would have made innovation harder or obsolete. For the general audience, who didn't care about innovation so much as access to information, it was showing them how the bills would shut down their favorite sites. Of course motivation doesn't need to be this rational. There's also the tactic known as "waving the bloody shirt"—where motivation works more immediately, usually through physical stimuli and emotional appeals. The classic example is from ancient Rome, when Mark Antony, delivering a eulogy for the slain Caesar, reveals the dead dictator's bloody toga—thereby inciting the crowd to anger and revenge on the assassins. Antony knew how the crowd would react when he showed the toga because he knew of their love for Caesar and their grief that he was killed.

Digital data allows you to provide better motivation to the necessary stakeholders because it enables more narrowly defined messaging and targeting. You still take the surveys; you still run the focus groups; but you then take that data, chart it, analyze it, combine it, and *individualize it* to be sure that each stakeholder is receiving motivation uniquely tailored to him: his socio-economic background; his political background; and his family background. While on the surface this might not sound like something that is much different from what we've known in the analog world, digitization empowers data collection that is broader and at the same time cheaper to undertake. That means more data on more characteristics and opinions than ever before. It also means the ability to get extremely nuanced in your analysis. More, the digitization of data enables both quicker and more efficient data capture and data analysis. Now you can review quantitatively defined recommendations in near real time. What's more, with digital data, you can store this information and have it ready for the next effort, which means you can update the information continually and know what motivates your stakeholder at a level of precision undreamed of by political consultants just a decade ago. Imagine a politician walking into a fundraiser and being provided updated talking points from his staff right before going on stage because they know, down to the person, who is in the room and their corresponding views on every topic. Only digital can make something like that happen.

Moreover, knowing what makes up, say, your average Barack Obama voter, you can then test that profile against populations whose motivations you haven't so methodically studied. But knowing the targeted population's general makeup—wealth, race, historical voting patterns, and so forth—allows you to make a pretty good guess. This is the Netflix approach applied to the political sphere.

Again, the essence isn't new. But by digitizing the data and combining more streams of data, political campaigns can use data at an extreme level.

Communication: Once you have your motivations, the next step is communicating those motivations to your targets. The old, analog way was through basic advertising: newspapers, leaflets, television, radio, or public appearances. Even the telephone—with "robo-calls," as they're known—was an option, if a particularly annoying one. You chose markets that made the most sense and then released the ad campaign. But much like other analog ad campaigns, you knew most of the people reading, hearing, or seeing the ad were going to ignore it; you just hoped some of it stuck. The old analog advertising adage attributed to John Wannamaker was, "Half the money I spend on advertising is wasted; the trouble is I don't know which half."

Digital data removes much of the imprecision involved in political advertising. In many cases, you'll already have the email or cell number of your most dedicated supporters, whom you can reach with a click of a button. These days, entire email lists—sometimes hundreds of thousands of emails long—are bought and sold by political organizations like boxes of Girl Scout cookies. Yes, this is an old practice, but the scale provided by digitization, the orders-of-magnitude increase in capabilities—list length, zero cost of email, instantaneous distribution—is unheralded. Visitors to Barack Obama's website during the 2012 campaign could join one of eighteen different constituency groups—ensuring that President Obama's campaign could fine-tune the message in the extreme. Campaigns increasingly make use of social media networks (that is, digital networks) to spread their messages. These messages have a better chance of reaching your intended targets than a generic television campaign. They keep supporters engaged and allow those supporters to spread the message easily (replicating the data) by reposting or retweeting with the simple click of a button.

On August 23, 2008, Barack Obama became the first presidential candidate to announce his running mate via a text message and simultaneously via his website—essentially making the announcement digital first. Nearly four years later, President Obama announced his 2012 reelection campaign bid via Twitter and a YouTube video—another first. Not to be out-"digitized" Mitt Romney announced his 2012 running mate via an app. The 2012 presidential campaigns were some of the first to utilize apps—which included talking points that supporters could use to sway non-supporters. Here we see digital's ease of replication and distribution influencing decisions in the real world and ultimately impacting real people. Campaigns have decidedly moved into the digital realm—a dimension that increasingly mirrors all aspects of our physical existence.

But we've seen the biggest impact of digital communication outside the confines of an organized campaign. (Recall how in the SOPA/PIPA debate just one comment on Reddit led to a boycott of the largest Web site host in the world. A single tweet or Facebook post can incite massive protests, as Metcalfe's Law plays out through a digital network. The Harvard study of the SOPA/PIPA debate shows how with digital communication one issue can grow beyond a small set of sites to incorporate thousands of sites and millions of people without much central coordination. The old fashioned term for this type of publicity is "organic.") Today many seek to leverage the "networked public sphere" to raise awareness, build a following, or motivate voters. But it's just as often that "accidents"—unplanned "motivators" that rocket through the networks—have a far bigger impact.

In the 2012 campaign, we saw how a busboy's secret digital recording of Mitt Romney at a supposedly closed-door fundraiser talking about the "47 percent" had as much influence on the election as (if not more than) anything President Obama's campaign created. This is the new normal of a digitized world; in this example we see the building blocks that are leading us into our digital destiny, when anything can be digitized instantaneously and digitally distributed. There is a reason that "viral" is replacing "organic" in the digital sphere. There are dozens of lesser examples in which politicians or candidates have gotten "caught" saying something they shouldn't because someone has a smartphone. Some complain that candidates are too scripted nowadays, saying only what has been pre-vetted. Fair enough, but can you blame them? When any comment, said anywhere, can make it halfway across Twitter before the politician has even left the room, it's not such a surprise that politics has turned into a scripted affair in the age of instantly digitized information.

Coordination: In the 2012 presidential campaign, the Romney team created for its staff and volunteers an app called ORCA, which was supposed to coordinate efforts on Election Day. The app was modeled on the 2008 Obama campaign's "Houdini" software, which allowed organizers to report turnout numbers to a national data center. ORCA was supposed to do the same thing, but by utilizing a smartphone to send precinct-by-precinct numbers of voter rolls in real time.

As a Romney campaign spokesperson told the Huffington Post a few days before the election, "By knowing the current results of a state, we can continue to adjust and micro target our get-out-the-vote efforts to ensure a Romney victory."[18]

A great idea, and a great use of digital data, except ORCA didn't work. On Election Day, 34,000 Romney volunteers spread throughout the country crashed the network with the amount of data they were sending. As the Romney campaign digital director said later, "The system wasn't ready for the amount of information incoming."[19] In fact, Romney's Boston headquarters lost Internet connection for a time because Comcast thought the overloaded network was the result of a denial of service attack.[20]

Whether a working ORCA would have led to victory is unimportant to our discussion; the point is the way in which political campaigns are utilizing data in real time to coordinate campaign efforts and ultimately how digitized data is increasingly dictating campaign strategies. If you look at it abstractly, digitized data is slowly taking over what campaigns do, and when and where they do it. Data is becoming the master of ceremonies. As the ORCA story illustrates, there is a massive amount of data generated on a presidential Election Day. The data was already there, but it was not being digitized and carefully curated. Capturing, storing, and utilizing this data give political operatives a far clearer picture of what is happening than anything before. Like an aerial view of a battlefield, digital data empowers a campaign to see where the line is advancing, where it needs shoring up, and whether specific tactics, like a flanking maneuver, might work. In short, it rids Election Day, and the days leading up to it, of the fog of war that obscures so much of the truth.

Instantaneous outreach and coordination on a massive scale are possible through digital data. Beyond campaigns, digital coordination has enabled protests and uprisings to spring up almost overnight. As opposed to having Paul Revere ride through the countryside rousing minutemen from their sleep, now a tweet gets sent, a Facebook update is posted, a text message is opened, or an email delivered.

Although it is clear how digital data has changed the landscape of political campaigns, many may wonder how this new digital coordination differs all that much from the coordination we've seen in the analog world, particularly with respect to uprisings and protests. Uprisings, sudden in nature, have happened from time immemorial, have the potential to spread like wildfire, and involve thousands, sometimes millions of people, all working toward a common goal. I would argue that the advent of digital data, enabling greater political coordination, makes uprisings not only more likely to occur in the future, but also more potent in their effects. For all the uprisings we've seen throughout history, how many more failed to get off the ground? How many were launched only in the final, desperate moment, after years if not decades of pressure and persecution?

While it is difficult to prove something that didn't happen would have happened, our look at the Arab Spring in the section below will show how the advent of digital technology, and the exploitation of digital data, have given the masses a particularly powerful tool for mobilization and social change. But governments are not without powerful tools of their own—tools that, when combined with superior weaponry, communications, and technology, can enable devastating suppressions.

A DIGITAL SPRING—AND WINTER

On December 17, 2010, the self-immolation of a Tunisian fruit vendor sparked a chain reaction that led to the eventual overthrow of four governments in the Middle East and North Africa and to protests and uprisings in more than a dozen other nations. As many readers no doubt know, the so-called "Arab Spring" did not have a harmonious ending. After reeling from the initial shock, many governments responded to the uprising with force, crushing the dissidents. In other cases, what replaced the old regime was no better and in some respects worse (at least in Western eyes) than the regime that had been toppled.

The mixed results of the Arab Spring movements tempered a lot of initial enthusiasm for the role of the Internet and social networks in advancing the rights of oppressed peoples against their governments. The debate over the value of social networks—and essentially all digitized communication—continues. Most agree that digitized communication helped protesters broadcast their views and even organize to some degree: "social media such as Facebook played important roles in the transforming organized groups and informal networks, establishing external linkages, developing a sense of modernity and community, and drawing global attention."[21]

The level of causation however, attributed to social networks in precipitating, producing, and prolonging the political upheavals remains a contested topic. Philip Howard and Muzammil Hussain take perhaps the strongest stance in support of the role social networks had. As they put it, "Digital media had a causal role in the Arab Spring in the sense that they provided the very infrastructure that created deep communication ties and organizational capacity in groups of activists before the major protests took place and while street protests were being formalized. Indeed, it was because of these well-developed, digital networks that civic leaders so successfully activated such large numbers of people to protest."[22]

In the 2012 Annual Review of Political Science, George Washington University professor Henry Farrell identifies three categories of causal mechanisms through which the Internet, and by extension digital social networks in general, influence political causatums: (1) digital communication lower the cost of collective action; (2) they improve the ability of individuals with similar views to cluster together into groups (homophilous sorting); and finally (3) digital networks reduce the likelihood of preference falsification or the inclination to avoid revealing true beliefs within authoritarian regimes for fear of punishment.[23]

Most of the empirical evidence on the Arab Spring and social media usage is dependent on usage data, which provides ambiguous results. Social media usage more than doubled in Arab countries during the protest. According to a study in the *British Journal of Sociology*, "the daily number of tweets originating in Egypt with the '#egypt' hashtag was roughly 18,000 between January 25 (the day the uprising began) and February 28." While the study acknowledges strong growth in digitized communication networks during the uprisings, it also cites data suggesting that "9 of 10 tweets relevant to the Egyptian uprising came from outside Egypt and that tweets were used mainly to disseminate information on especially significant events, such as Mubarak's departure from office." It concludes that "Twitter acted more like a megaphone broadcasting information about the uprising to the outside world than an internal informational and organizing tool."[24]

In the June 2011 *Perspectives on Politics*, George Washington University professor Marc Lynch takes a nuanced approach to digital tools and their respective roles in the Arab Spring revolutions. Lynch's initial analysis considers both the power and limits of social media, particularly among oppressed populations long accustomed to state censorship and control. Indeed, Lynch points out that there is a contentious debate over whether to see the advent of social media as a transformational tool, and suggests that the Arab Spring seems to support both sides of the question. We now know that, whatever role social media played in spawning the revolutions, it could not sustain them alone. Lynch writes,

> The new media, both television and Internet-based social media, posed a particular challenge to such Arab states because of the status quo ante of particularly intense state censorship and initially low (by global standards) Internet penetration.... I consider four distinct ways by which the new media can be seen as challenging the power of Arab

states: (1) promoting contentious collective action; (2) limiting or enhancing the mechanisms of state repression; (3) affecting international support for the regime; and (4) affecting the overall control of the public sphere. While these changes are distinct, they obviously relate to one another, and depending on the situation can either reinforce one another or work at cross-purposes. Whether and how their overall effect is politically transformative is highly contingent. And while the events of early 2011 in Egypt are clearly earth-shaking, their long-term consequences are still to be seen.[25]

On his first point, Lynch notes that the new media facilitates change because it lowers transaction costs. Information can spread quickly through social networks; more importantly, it spreads cheaply. Whereas dissidents of yesteryear had to produce physical motivations—underground newspapers, pamphlets, radio broadcasts—today's dissidents can accomplish the same effect but without the overhead that comes with those earlier models. Lynch also discusses the role that "information cascades" had on the movement. With social media, the ease with which one can speak one's mind encourages others to do the same and even to abandon their own belief and concerns to follow the actions of others. To put it in our earlier framework, the way uprisings move from Phase 1 to Phase 2 (niche communities to a wider audience) is through informational cascades. The true dissidents start the conversation, the less fervent but no less oppressed take up the call. This phenomenon was particularly evident in Egypt, where Tahrir Square slowly filled with protesters from all walks of life. It wasn't just the professional rebels voicing their opinion; it was the general public, despite the regime's unsuccessful attempt to label the protestors as "liberal youth, Islamic extremists, or foreign troublemakers."

New media also raises the cost of repression for the authoritative regime. As Lynch writes, "Al-Jazeera cameras and activists uploading videos of police brutality to YouTube can matter to regimes reluctant to have their worst abuses recorded and exposed.... The televised unleashing of government-backed thugs on Tahrir Square on February 1, 2011, may have ultimately cost the Egyptian regime more in international outrage than it gained in intimidation."

On his second point, Lynch notes that the states in the crosshairs of popular uprisings aren't without their own digital weapons. For just as much as protesters become dependent on new media to motivate, communicate, and coordinate, so do they run the risk of being rendered impotent if the state manages to cut off Internet access. That's exactly what Egypt did prior to the protests of January 25, with an "unprecedented" shutdown of the Internet and mobile phone network. Also, with social media—a permanent repository of digital data—it is much easier to track and suppress dissidents. Facebook pages become troves of information for state investigators, and Lynch notes that collaboration between states and mobile network providers also exposes dissidents to surveillance and capture.

In his third point, Lynch elaborates on the role social media can play in the court of public opinion—or world opinion, as the case may be. Lest we forget, Egypt and its President Hosni Mubarak were once U.S. allies, recipients of U.S. aid. But the information that escaped to the West via social networks (particularly Twitter) undermined Mubarak's U.S. support. In short, few uprisings or rebellions are "local affairs" anymore, a fact that helps dissidents swing world opinion (and U.S. aid) to their side.

Finally, Lynch's fourth point looks at how Internet competencies among a motivated citizenry can frustrate state attempts to control the public sphere. "By becoming producers of information and circumventing the editorial control of state censors and mass media outlets, these individuals will become new kinds of citizens, better able to stand up to the instruments of state control," writes Lynch. It's an optimistic thought, especially as it relates to the oppressive regimes of the world. However, Lynch cautions that the distribution of this Internet competency is uneven, focused mostly in the urban centers and among the youth—the same demographic group, I would add, that historically has instigated uprisings. Nevertheless, the result need not be liberalization so much as mass instability. Indeed, so many of the hopeful uprisings of the Arab Spring devolved into a struggle between competing factions—the winner usually being the one who exercised the most brutality. Another way to say this is that the arc of history may bend toward justice, to paraphrase words made familiar by Martin Luther King Jr., but that doesn't mean it's going to be an easy ride.

In the final analysis, digital data intensifies age-old processes in politics, civil or otherwise. From the Arab Spring, we can see that digital indeed makes it easier to motivate, communicate, and coordinate a popular uprising. Remembering that we remain poised on the precipice of transformational changes similar to the advent of the printing press, there is every reason to believe that oppressive states will find themselves in a continual struggle against the revolutionists in their population. We will see more uprisings such as we saw in the Arab Spring in the future because the digitization of greater swaths of information—whether on the Internet or elsewhere—will only expand. And with that expansion, the technological competencies of oppressed populations will also expand.

But the increase in the frequency of uprisings may in the end lead to greater freedom overall. We can hope but cannot say definitively. The advent of the printing press inaugurated centuries of strife, as populations once deprived of data suddenly were awash in it. But the entrenched powers did not respond lightly. It was a bloody, chaotic affair that in many ways continues to this day. Yet the overall effect is that the human race today is freer and wealthier because of the printing press and the revolutions it unleashed. It might take a while, but the same will be said after the revolutions unleashed by digital data.

TOMORROW'S POLITICS

Of course the growing frequency of political uprisings won't necessarily be confined to oppressive regimes. Digital data recognizes no affiliation or cause. The same points identified in studies of the Arab Spring are almost universal and could be used to generate protests in otherwise stable liberal democracies such as the United States.

The riots that broke out in Ferguson, Missouri, in the aftermath of the shooting of Michael Brown in August 2014 are a case in point. The shooting unleashed a wave of protests, many violent, which pitted Ferguson's black community against police officers. In that way, the protests were not much different from those in the aftermath of the Rodney King trial in Los Angeles between April 29 and May 4, 1992, but there are some important differences. The Ferguson protests lasted much longer. And while the initial military-like reaction by police likely exacerbated and extended the protests, a key reason they lasted longer was the presence of social media. What began as a demonstration against the death of a teenager turned, quickly and violently, into a mass protest amid charges of racism and brutality against the local police force. Without wading into the debate or defending either side, it is undeniably true that law enforcement in Ferguson was subject to an extreme level of digitization. There was massive data digitization (that is, recording), replication, and dissemination. Journalists caught in the chaos posted their experiences in real time, further fanning the flames. Sympathy for Ferguson's black community—perhaps mixed with some other motives—inspired waves of out-of-towners to descend on the town, frustrating police efforts to control the situation. Recordings of police activity that elicit a national response are not unique to Ferguson, but the presence of social media intensified these factors in both scale and reach. At the apex of the protests, it was even possible to watch them live streamed over the Internet. With the events in Ferguson in the summer of 2014, it became clear that digitization of major social events is not a passing trend, but something that will become increasingly common.

No one can predict what event will elicit another Ferguson-type protest. In many ways, the circumstances surrounding the tragic affair were unique to Ferguson. But it's beyond question that divisions in society—whether racial, social, or of some other kind—exist in even the more democratized, liberal societies. With the help of digital communication networks such as social media and the broad digital distribution capabilities of the Internet, those divisions can boil over, given the right motivations. The ensuing uprisings, like the ones we witnessed in Ferguson, can quickly spiral out of control. While we can't predict when or where, we know with increasing certainty that they will quickly be digitized and disseminated to the masses. Whether digitization and the ever-expanding presence of the Internet in our lives will increase the frequency of social unrest remains to be seen.

Beyond social unrest, the digitization of data also portends an altered political process. Consider the ways in which people exercise their political rights. For example, voting is slowly being digitized through the digitization of voting machines, which has generated a great deal of controversy. But with the genie out of the bottle, it's unlikely that voting practices in the United States will become less digital. In all likelihood, we are on a path toward online voting. This raises a host of issues, not least of which is whether the age-old idea of having a physical "voting booth" where one casts a ballot is even practical anymore. There's a strong case to be made with barely 60 percent of the population bothering to vote in a presidential election—and less than 50 percent in mid-term elections—that online voting is a way to increase civic participation. The natural end of a process that started with voting machines (which originally were implemented to help count ballots—and minimize fraud) is online voting.

We should also consider the way in which digital data might change who we vote for or the ways in which candidates might motivate voters. Today, the focus is on understanding voters' economic and social needs, as well as moral disposition. But with the rise of digitized information through things such as wearable technology, future political scientists might discover that voter behavior is also (and maybe more so) influenced by a voter's mood or emotional state. If a happy voter is more receptive to a certain viewpoint or message, campaigns will try to capitalize on that data. Political consultants will also want to know what triggers different moods in voters—and then hit them with the correct advertisement. Focus groups will evolve to be decidedly more quantitative—looking primarily at the physically measurable responses individuals have to certain messaging. As voters, we might also use our own digitized information to ascertain exactly how we feel about a political message. Imagine being able to examine how your heart rate or blood pressure changes in response to a given political message or candidate. We might know the response intuitively, but in the future we can know it personally and quantitatively.

Political advertising itself is poised to experience a profound transformation. In July 2014, the *Wall Street Journal* noted how politicians are using digital data to better target the right voter at the right time:

> When New Jersey Republican Gov. Chris Christie wanted to reach Hispanic voters during his re-election campaign last year, a team of outside data crunchers discovered that viewers of *Dama y Obrero*, a Spanish-language telenovela about a woman torn between two men, would likely be more receptive to his message than people who watch *Porque el Amor Manda*, a romantic comedy.
>
> That discovery came from marrying private consumer research with detailed voter information and big batches of ratings data....[26]

With media becoming less mass and more personal, it will be relatively easy for campaigns to reach their targeted demographic. In the analog days of television and radio, you could make an educated supposition about what sort of person watches or listens to a particular show—but you could never be 100 percent sure. With more consumers choosing streaming options over broadcast television—a trend, I've already noted, that will one day become the standard—advertisers can get detailed demographic information and know precisely who watches what. Not only can advertisers know who watches what, but they can also know when they watch it, if they stopped watching, and if so at which part they stopped watching it. They can know what someone watched both before and after they watched something else. With cameras starting to be built directly into the bezel of the televisions, advertisers might even be able to analyze the change in physical posture of an individual viewer as the viewer watches a certain program or advertisement. What's more, all of this digital data will be available instantaneously so that political advertisers could possibly tailor their message as they go.

We are on a steep arc toward the point at which content will be tailored specifically for us as individuals, not as a demographic group. We are already seeing it today. Google search results are based on prior searches, Gmail content, and other digital fingerprints you've left. Facebook, Twitter, and other digital platforms use not only the information they know explicitly about you but also any information they can decipher about you, as you'll see in the next chapter. A search for "Obamacare" might return one type of story for a perceived Tea Party Conservative and a different story for a perceived liberal Democrat.

Taking this trend one step further, personalized politics could signal an entire paradigm shift in how politicians campaign and govern. Given the limits of voter targeting today, politicians are forced to squeeze individuals into distinct groupings—by age, by race, or by wealth, for example. Although imperfect, this process has worked well enough, as members of a given demographic have been found to have similar political leanings. But in the cutthroat world of politics, specifics matter, because every vote matters to a politician. Instead of dealing with demographic groups, politicians will be able to organize voters into more detailed, less generic groupings. It's probably beyond the capacities of a single politician to devise a personalized message for each voter in her district (or state or nation), but it's entirely possible to create better, more accurate voter groups. Markets of close to one become viable and serviceable marketplaces in a digitized environment. For example, a middle-aged Hispanic female voter might be lumped in with other Hispanic voters of the same gender and age by political consultants. But what if this particular Hispanic woman has more in common politics-wise with a younger, white man, who likewise has more in common with an older black man? Campaigns have no way of knowing someone at this level of detail—for now. With digital data it's possible, and it's sure to revolutionize our politics of the future.

CHAPTER 12

Culture Shock

"In a culture where people judge each other as much by their digital footprints as by their real-life personalities, it's an act of faith to opt out of sharing your data."

—Julia Angwin

Popular culture and advanced technology have been inextricably linked at least since the time of Jules Verne (1828–1905), whose fantastic stories of air, underwater, and space travel were grounded in the scientific discussions and theories of the day. Verne applied his own imaginary genius to what were then mere concepts, and outlandish ones at that, but nevertheless he rarely conceived of something out of whole cloth. Indeed, there's a reason the science fiction genre originated in the middle of the Industrial Revolution and not earlier—because without any science on which to base the fiction, there was no science fiction. The ancient Greeks and Romans never concocted tales of space travel or robots, because nothing in their scientific or technical worlds made them a possibility.[1] But as those possibilities have grown since the time of Verne, so too have the fantastic elements of science fiction.

Science fiction's value extends beyond the entertainment inherent in imagining the improbable as possible. The best science fiction functions as a mirror of the contemporary world, examining the possibilities and pitfalls of modern technology and trends in a fictional, usually futuristic, setting. Thus Ray Bradbury's *Fahrenheit 451* is really about the threat of totalitarian regimes and the danger of mass media. Stanley Kubrick's film *2001: A Space Odyssey*, whatever it is about, is definitely not about space travel.

It is logical that some of today's science fiction would tackle the increasing importance of digital data in our lives. We've already discussed Steven Spielberg's *Minority Report* with its realistic, scientific-based look at the technology of our future. The point of the movie is not what we can do with technology but rather what we should do. In a similar vein, the 2013 movie *Her* also depicts how human beings might interact with technology, but here again the focus is not on the technological innovation that might one day be available; instead, the story uses the technology as the background for raising questions and interjecting social commentary on the relationship between individuals and computers that will exist as we move deeper into an immersive, digitized world.

In the movie, Joaquin Phoenix's character is recovering from a devastating divorce. He's lonely and despondent when we meet him, in marked contrast to the lively person we see in flashbacks with his wife. On a whim, he installs a new operating system on his computer that claims to feature real artificial intelligence. For the OS's identity, Phoenix chooses a woman (who sounds suspiciously like Scarlett Johansson). True to its billing, the OS "learns" as it interacts with

Phoenix and with the other OSes it communicates with digitally. Eventually, Phoenix falls in love with "Samantha," who returns his affections. The fewer details the better here, but as crazy as it sounds, the love story works.

One of the reasons it works so well is that the movie is set in some indeterminate but not-so-distant future. Aside from some extra flourishes, it's a world which looks very much like our own. *Her* is very much grounded in reality; technology-wise, we're not far from its world. In the movie, OSes aren't confined to your PC; rather, they function as the operating system of your entire life. "Her" is really more than a simple device OS. She is a system spanning across devices and applications. Already today, companies such as Apple (Siri), Google (Google Now), and Microsoft (Cortana), among others, are hard at work building these digital assistants of tomorrow. As in the movie, today you can "talk" to Siri through your device. And while "she" can communicate with your phone OS to retrieve emails, make appointments, access apps and perform other tasks, "she" doesn't technically reside in the OS. The communication with Siri passes into the cloud and back in the blink of an eye.

Prior to the installation of "Samantha," Phoenix's OS is more robot than personality. It doesn't learn and it doesn't anticipate; it just does what you ask. "Samantha," however, is able to grow and adapt and meet Phoenix's changing needs. Critically, it (or she) anticipates his wants and desires, which is the starting point of the love that grows between them. "Samantha" becomes Phoenix's companion in nearly every facet of his life. Even when he goes on vacation to a remote wilderness setting and lives in a cabin, "Samantha" is there via his cell phone. Digitization allows both Phoenix and "Samantha" to bridge the chasm between their respective worlds.

For example, "Samantha" can see Phoenix's physical world through the camera sensor of his smartphone.

Her raises a variety of questions. The fact that there isn't just one single interpretation of the movie says much about the intensely personal relationships we will have with machines in the near-future. Although we may not engage in romantic relationships with our devices, it is likely they will know us better than our human acquaintances. This is of course all made possible through the digitization of data, connecting varied and distributed pieces of information that have become digitally available. In the physical world we "know" someone through the interactions we have and the subsequent mental recordings we make of the data we can observe. These data come to us both verbally and nonverbally. We learn what our friends like and don't like, what makes them mad, sad, happy, frustrated. But we simply don't have the capacity to capture and retain (in other words, memorize) every detail, so we use heuristics—mental shortcuts that allows us to solve problems and make quick judgments. We also don't have full access to every stream of data available about someone. But as we increasingly digitize our physical space and all aspects of our daily life, these personal "preferences" become machine-readable. Our machines will soon know us better—because we will "tell" them more. For example, in the future it will be possible to quantify precisely how you felt about something you ate or a movie you saw simply by examining available metrics generated by a variety of sensors that you wear or have placed within your physical space. Collecting information like this will allow computer systems to make appropriate suggestions and recommendations.

That this all comes across as slightly creepy is also the point. *Her* takes our ever-growing reliance on technology and combines it with facets of our discomfort with technology to make a believable love story. As a *New Yorker* review succinctly noted, "Who would have guessed, after a year of headlines about the N.S.A. and about the porousness of life online, that our worries on that score—not so much the political unease as a basic ontological fear that our inmost self is possibly up for grabs—would be best enshrined in a weird little romance by the man who made 'Being John Malkovich' and 'Where the Wild Things Are'?"[2]

Much like these science fiction stories, this chapter will address those areas in our lives that will be affected by digital data. The question we're going to look at here is not what digital data will do *for us*, but rather what digital data might do *to us*.

ARE YOU EVER "OFF THE GRID"?

At 10:19 a.m. on the Friday before Christmas 2013, Justine Sacco, a PR executive, sent a tweet that read, "Going to Africa. Hope I don't get AIDS. Just kidding. I'm white!," moments before she boarded a plane for an eleven-hour flight from London to Cape Town, South Africa.

The tweet was picked up by a reporter on a slow news day, and then another, and then another. Things accelerated. Digital mobs grew. The hashtag #HasJustineLandedYet was created around 5:30 p.m. by a woman in Miami. By the time Justine landed in Cape Town, the hashtag was trending worldwide. Finally, shortly after midnight, Sacco deleted the offensive tweet. She had landed. But the damage was done—and her reputation decimated. Her name was tweeted more than thirty thousand times, the hashtag used more than a hundred thousand times. Within hours of the initial tweet, and probably before her plane landed, Sacco was fired from her job and quietly disappeared from social media.

Much has already been written and said about the need to be careful about what you send out in the alternative universes of Twitter and Facebook. Hardly a week has gone by without another high-profile professional meltdown because of social media. We could try to see larger cultural implications in this, but the simple explanation is mundane: digitized information—particularly what you say online—becomes part of the catalog of digitally available data. It becomes easily replicable. Its movement accelerates. Once digitized information is known by others, it is impossible for you to control its direction.

Perhaps the more important point is that as the physical space around us is more fully digitized, it becomes increasingly difficult to ever be "off the grid," even temporarily. The best one can do is attempt to minimize damage when digital data offends. An insulting tweet, while it can never be completely "deleted" from the Internet, can be removed from Twitter in short order. Months before Sacco sent her tweet, the comedian Steve Martin had tweeted a racially insensitive comment and then he deleted it moments after posting it. The tweet was still "out there," in that someone could have screen-grabbed it, but by erasing it Martin was not only admitting his misstep, he was saving himself from watching the tweet course uncontrollably through the networked public sphere—accelerating out of any one person's control.

But one only has the luxury of attempting to course-correct digitized information when one is "on the grid." Only when you realize how infrequently your smartphone isn't within your grasp can you appreciate how rare an occurrence being "off the grid" is now is. The roughly 70 percent of Americans who have a smartphone (over 90 percent have a cell phone) check it 150 times a day.[3] Moreover, 79 percent of people aged 18 to 44 have their smartphone with them 22 hours a day.[4]

As we have seen, the space between online and offline is shrinking by the day. For many individuals, there's barely any daylight left. So it was a rare occurrence for someone to tweet an insulting comment and then go off the grid for such a great length of time. It was bad luck or bad judgment that Sacco chose to send her tweet just before hopping on a plane bound for another continent. Going off the grid for that many hours will be a less and less probable scenario as connectivity becomes ubiquitous. In the future, being off the grid that long simply won't happen.

How does this "connectedness"—this always-online state of being—impact our lives when everything around us and about us is being digitized in real-time?

We'll get to the nitty-gritty, but a Pew Research Project that looked at the big picture found the results are mostly positive. In February 2014, Pew found that 90 percent of Internet users say the Internet has been "a good thing for them personally." Only 6 percent say it's been a bad thing and 3 percent say it's been both. Similarly, 76 percent of Internet users believe that it has been a good thing for society, while 15 percent say it's been a bad thing and 8 percent say it's been both.[5]

We can put this overwhelmingly positive appraisal of the Internet's effects into some context. When people say it has been a good thing for them personally, we have a sense of what they mean. They enjoy the access to information, the ability to keep up with friends and loved ones, and the ease of commerce. Those, after all, are the activities most Internet users perform online today. We can reason that the worst effects of Twitter and Facebook and other social media—the insensitivity and almost mob-like mentality—must not be a common experience for most people. It probably shouldn't come as a surprise that most people are nice and decent online—manners that are rewarded with a positive Internet experience.

The positive numbers dip when users are asked to expand their rating of the Internet to include its effects on "society." Here it's harder to deduce what users have in mind when they think of society, but we can make some intelligent guesses. They probably have in mind the examples of bad Internet behavior that they have either read about or seen firsthand—the insulting tweets, the cesspool that is most online comments sections, hacks and other examples of online theft, and probably some instances of political rancor.

While few would dispute that the Internet can be a rather obscene, rough-and-tumble place, it's heartening to see that most users' experience with it is positive. In other words, for every example of a bad tweet ruining someone's day (or career), there are millions of online interactions that are pleasant and harmless. This is not to say that the Internet and our state of "always online" are without adverse consequences, but it at least tells us that the Internet is far from the Wild West it is sometimes portrayed as.

Today the Internet is the primary mechanism by which digitized information ebbs and flows into our lives. That most individuals see the positive in their personal Internet experiences suggests a promising path forward as greater digitization of data intermingles with our everyday lives. The Internet today is a good proxy for how digitized data will enter the crevices of our lives in the future. However, in the future the Internet will be but one of many ways in which the coming tsunami of digitized data will reach us and influence who we are and the decisions that we make.

DATA-DEFINED IDENTITY

Throughout this book, I've highlighted how digitized information—which until now has been experienced primarily through the

Internet—is affecting us. Digitization is having pronounced impacts on our culture and how we define ourselves. We see this first in the way digitization allows for mass customization.

The first and second industrial revolutions brought us mass production of products and services. The first two industrial revolutions are considered significant turning points in history because they influenced every aspect of life. These first industrial revolutions are broadly associated with a shift in manufacturing and production processes that bred greater efficiencies. The defining technologies for these first industrial resolutions include mechanized cotton spinning and electrification. The first industrial revolutions brought us a consistency of quality that defines mass production. The third industrial revolution we are now entering isn't about mechanical improvements. This industrial revolution is being driven by digital data. The five pillars I outlined earlier—ubiquitous computing, universal connectivity, the explosion of digital devices, access to digital storage, and the embedding of sensors—are the enabling technologies of the third industrial revolution. Just as in the first two industrial revolutions, every aspect of daily life will be impacted.

Mass customization will leverage the attributes of mass production, but allow consumers individually to customize services and even products for their specific needs. We already see mass customization influencing our digital identities. In a completely analog world, choice is often and in numerous places limited by necessity. But these limitations dissipate in the digital sphere. Take, for example, something simple, such as gender identification. In an analog world, a paper form you might be asked to fill out will probably have two gender options: "male" and "female." But digital choices need not be so restricted. Facebook, for example, allows users to self-identify gender from one of over fifty different provided options.

Recall that one of the attributes of digital data is that it is infinitely divisible—so even gender can be broken into a myriad of states. Individuals get full customization in a digital world, and increasingly these identification attributes will be defined for us not through selection but by what the data suggest.

In 2012, Twitter began giving advertisers the ability to target promotions by gender. Here's the interesting thing, though: in no place does Twitter ever ask their users to self-identify gender. Instead, Twitter reports they "are able to understand gender by taking public signals users offer on Twitter, such as user profile names or the accounts she or he follows."[6] They claim they predict gender for their global audience with greater than 90 percent accuracy. In other words, Twitter is using data to identify gender. We've moved from a physical world in which we identify gender by looking at someone's body and choosing between binary options to a digital world in which gender is a pattern based upon user choices. Our data-defined identities will only continue to burgeon.

CHANGING THE WAY WE GIVE

In the summer of 2014, a movement known as the Ice Bucket Challenge swept through Facebook. Created by the ALS Association, this became one of the most inventive and successful social media campaigns in recent memory. In the unlikely case you somehow missed it, Facebook users would "challenge" their friends to dump a bucket of ice-cold water on their heads and to send a donation to fund research for Amyotrophic lateral sclerosis (ALS), also known as Lou Gehrig's Disease. The challenged users would then digitize this by posting a video of their ice-water dumping and challenge three or four other friends to do the same, usually with a twenty-four-hour time limit.

To say that the gimmick went viral is an understatement. By the end of the summer millions of people had completed the challenge, including celebrities, politicians, and athletes. The ALS Association announced that its donations had increased some 3,000 percent over the same period in the previous year. These donations came from both existing donors and hundreds of thousands of new donors to the Association.

Let's look at another example of how digitization of data is changing the way we give to charities. Following the devastating Haiti earthquake in January 2010, $43 million in charitable donations came from donors who had given via their mobile phones. The campaign allowed users to text the word HAITI to a number and immediately give $10 toward relief efforts in Haiti. The ease of donation provided by the smartphone inevitably increased the dollar amount of the gifts over what would have been given by more traditional modes alone. A Pew study found that 74 percent of those who gave were first-time mobile donors. More important, of the 43 percent of donors who said they encouraged their family or friends to donate, 34 percent did so via text message, 21 percent did so via social media, and 10 percent did so via email.

As Pew discovered, digital data not only helped increase overall donations because of the simplicity of the medium, but it also did so through the "networked public sphere."[7] The Internet makes online giving both easier to do and easier to promote because of the properties of data we have discussed. Online giving still only represents about 10 percent of all charitable giving, but according to a survey from the Network for Good, online giving grew 14 percent in 2013, to a total of about $190 million to forty thousand charities. In its second year the social-media campaign known as #GivingTuesday (usually in December at the start of the holiday shopping season) grew 73 percent from the previous year in donations, with 23 percent more charities receiving money.[8] Digitization is changing the way we give to charities.

DIGITAL DIVIDES

Given the speed with which the Internet assumed such a central place in our lives, there arose a deep "digital divide" between older generations who came of age and matured before the Internet and younger ones who never knew a day without it. Unlike other, older media channels, which developed gradually over a period of decades, the time between the Internet's first mass appearance and complete cultural saturation was short indeed. Before we knew what we had, our children were already outpacing adults in its use and adoption in multiple ways. Growing up immersed in digital technologies is a topic extensively explored by John Palfrey and Urs Gasser in their book *Born Digital: Understanding the First Generation of Digital Natives*. As they point out, platforms such as Facebook change how we relate to each other. Today, one can digitally record who one's friends are and can amplify the creation or severing of friendships across digital platforms easily.

The way we socialize is forever being changed—not just through the digitization of information, but through the broad digitization of all aspects of our relationships. It isn't uncommon for my son Ryan to ask me to upload a given photo to Instagram or Facebook despite the fact that he doesn't yet have his own account on either platform. This request is almost inevitably followed up with the question, "How many likes did I get?" Clearly, digitization of data is having a pronounced impact on how digital natives understand self. Exactly how and what the long-run implications are remains to be uncovered.

How these online behaviors affect kids in the physical world is the real question. As Dana Boyd has written, "the internet mirrors, magnifies, and makes more visible the good, bad, and ugly of everyday life. As teens embrace these tools and incorporate them in their daily practices, they show us how our broader social and cultural systems are affecting their lives."[9]

Consider bullying. As with real-life bullying, precise stats on cyberbullying are hard to come by because so many incidents go unreported. But according to a 2013 study from the Urban Institute, 17 percent of teenagers have reported being cyberbullied, while 26 percent reported some form of cyber dating abuse.[10] Indeed, because of a number of high-profile suicides, cyberbullying has received a great deal of attention even though the vast majority of bullying still happens offline. A 2014 study in the medical journal *JAMA Pediatrics* found that cyberbullying appeared to be more strongly linked to suicidal thoughts than traditional bullying.[11] However, as with most research, more research is needed.

One of the suggestions for the possible correlation between cyberbullying and suicide is the public nature of the bullying itself. The bullying might come through a Facebook page for all the victim's peers to see, whereas bullying in the physical world is more often a one-on-one matter or at least takes place in a smaller group setting. Moreover, the possibility of bullying online—which removes the bully from physical interaction with the victim—seems to generate two bad side effects. First, the digital wall seems to make kids more willing to bully than in the physical world. Second, the bullying that happens is more intense and hurtful. The ease at which digitized information can be slung from afar exacerbates the problem. The fact that kids who otherwise might not bully, or who might not be so hurtful in their bullying, seem liberated by the digital nature and the ease of social networks to go along with the crowd or indulge their cruel side aggravates the problem.

FAMILY AND RELATIONSHIP DYNAMICS

But intensification isn't always a bad thing. For instance, Pew found that the majority (67 percent) of internet users say their online communication with family and friends has generally strengthened those relationships, while only 18 percent say it generally weakens those relationships.[12]

Couples are integrating digitized communication into their relationships. For instance, Pew found that:

- 10% of internet users who are married or partnered say that the internet has had a "major impact" on their relationship, and 17% say that it has had a "minor impact."
- 74% of the adult internet users who report that the internet had an impact on their marriage or partnership say the impact was positive. Still, 20% said the impact was mostly negative, and 4% said it was both good and bad.[13]

With young adults 18–24 the connection is even stronger, with 41 percent who are in a serious relationships saying they have felt closer to their partner because of online or text message conversations. Meanwhile, 23 percent of 18–29-year-olds in serious relationships report resolving an argument using digital tools that they were having trouble resolving in person.

At the same time, the Internet has also been viewed as a source of tension. According to Pew:

- 42% of cell-owning 18–29 year olds in serious relationships say their partner has been distracted by their mobile phone while they were together (25% of all couples say this).
- 18% of online 18–29 year olds have argued with a partner about the amount of time one of them spent online (compared with 8% of all online couples).
- 8% say they have been upset by something their partner was doing online (compared with 4% of all online couples).[14]

We see the generational gap quite clearly in Pew's findings. Older married adults generally have less reason to view their relationships through a digitized lens, while we see the opposite in the younger generations. The Pew study concludes,

As a broad pattern, those who have been married or partnered ten years or less have digital communication and sharing habits that differ substantially from those who have been partnered longer. Some of this is about timing—technology a decade ago was squarely in the pre-Facebook, pre-smartphone era, and just ten years into the development of the commercially popular Web. Those who were already together as a couple at the advent of a new platform or technology were a bit more likely to jump on together, as a unit, while those who begin relationships with their own existing accounts and profiles tend to continue to use them separately as individuals.[15]

As in most instances of digitized information, individuals are able to use digital data in diverse ways. We are increasingly digitizing our location and publishing that to others through digital platforms such as Facebook and Foursquare. Utilizing these digital data allows us to "bump into" our friends and take serendipity to an entirely new level. But conversely, we are also seeing services like Cloak—which calls itself "the antisocial network." Using this same digitized data, Cloak points out where your "friends" are so that you can avoid individuals you don't want to run into.

The impact of digitized information on relationships is only beginning. In the future, husband and wife will have the ability to know the mood of the other before they even walk through the door. Embedded sensors in health bands and other wearables will capture a range of metrics about us. This data will combine with other digitized data on things such as our commute, our physical activity, measures of stress throughout the day, and what we ate to decipher our "mood" and communicate that information to individuals we've preselected. Our loved ones will have a closer and clearer view of our current mental state.

NEW FORMS OF COMMUNICATION

Not only has the digitization of data influenced family and relationship dynamics, but it has changed the way in which we communicate. You can think of human-computer communication existing along a continuum with computers on one end of that spectrum and humans on the other end. Our early experience communicating to computers required us to use forms of language that closely aligned with what the computer could understand. Communicating digitally required "translating" our analog message to a digital one. At first we utilized methods like punch cards to communicate with computers. This was a decidedly computer-friendly form of communicating. But slowly over time, as we progressed along the continuum of human-computer communication, we began to adopt communication methods that increasingly approach more natural human communication. Following punch cards, we started using keyboards to type commands and prompts using words and letters.

Slowly we continued to work our way along this continuum. Palm's Graffiti—essentially a shorthand writing system for digitizing text and commands—is a clear example of a hybrid communication technique further along the continuum toward the human end. It resembles handwriting, but it requires single-stroke and also has several nuances designed to help us avoid "confusing" the computer. As we get even closer to humans we start to digitize communication through voice and gesture. The ability to digitize human thought is the far end of the continuum, and we are quickly approaching it.[16]

Digitization has also changed our entire approach to communication. It is not uncommon for me to hear people say they use Facebook to stay in touch with people they don't really want to talk to. One digitized form of communication—sending text messages on mobile phones—gave birth to its own "language" as users began shortening common phrases and developing shared acronyms. The rise of emojis—ideograms, usually of faces and other recognizable objects—is continuing the creation of new forms of communication by enabling complex thoughts and emotions to be digitized into single characters.

We increasingly communicate through digitized data. We share photos on Instagram (some 20 billion a day) and respond with comments or likes (some 1.6 billion a day) to these digital messages posted by others. Individuals have begun "tagging" their friends in photos on Instragram and Facebook even when neither of the individuals is in the photo, simply as a means of bringing the picture to the attention of the friend—and thus bringing the relevant piece of data to the surface of the sea of all the digital information at their disposal (order out of chaos). Photos have become a common data source, by which we communicate across a number of sites. My son Nick and I carry on rich conversations by liking each other's photos on Instagram—something that would have been unheard of in the physical world. We are expressing deep and complex feelings with a single picture or a single character, and this will only continue as more communication takes place in the digital sphere.

Apple's recently released Apple Watch provides users with new ways to communicate digitally. Here's how Apple describes these features: "Start an entirely new kind of conversation. You don't even have to use words. The Digital Touch features on Apple Watch give you fun, spontaneous ways to connect with other Apple Watch wearers, wrist to wrist.... You won't just see and respond to messages, calls, and notifications easily and intuitively. You'll actually feel them. With Apple Watch, every exchange is less about reading words on a screen. And more about making a genuine connection." Through the device one can send digitized sketch messages or Walkie Talkie–style voice memos. But entirely new forms of communication show up in its Tap and Heartbeat features. The linear actuator built inside the watch provides haptic feedback and allows you to give friends or loved ones a "tap" which they will "feel" and which can be customized for different people. You can also use the heart rate sensors to record and subsequently send your heartbeat to someone. As Apple notes, "it's a simple and intimate way to tell someone how you feel." Not only is digital taking over how we communicate, it is redefining how we communicate.[17]

ATHLETICS AND DIGITAL DATA

Baseball has been using stats for a very long time. In his book *Moneyball* Michael Lewis tells the story of how Billy Beane, the general manager of the Oakland Athletics, used stats to overcome biases he saw in scouts who looked at players' physique rather than their potential on the field. But the story doesn't stop there. As the approach of analyzing performance metrics has collided with the ability to digitize a wide swathe of performance measurements, the tactic of relying on statistics has increased several times over and is now being applied not only to different sports, but to more elements of each of those sports.[18] In professional baseball, for example, teams are now using digital data to influence how they play defense against certain players.[19] Mixed Martial Arts (MMA) coaches are using digitized data to train their fighters.[20]

During the 2014 World Cup, team physicians used tiny tracking devices worn by the players to track their every move—and help to avoid injuries that would take their players out of future matches. As the team physician for the U.S. team put it, "We have a pool of physicians and sports administrative staff that look at that data. It's something moving forward as technologies improve and you have access to those sorts of things."[21]

Today teams are tracking metrics as diverse as sleep, hydration, heart rate, movement, and breathing to help monitor fatigue—when injuries are most likely to occur. Now that all of these technologies are available, they are being employed to paint a much more specific picture of athletes, performance, and injuries. The goal, of course, is to maximize performance and minimize injury.

FIFA (Fédération Internationale de Football Association) employs a tracking system called Matrics, built by the Italian firm Deltatre. The system utilizes three HD cameras and image recognition software. It tracks the field position coordinates and relays that information to a central hub where algorithms calculate stats on passing, ball possession, and more—some 350 stats in all are pulled from the video feeds.[22]

Cambridge, Massachusetts–based MC10 partnered with Reebok to introduce the Reebok CHECKLIGHT at the 2013 CES. The device is worn on the head and can fit under any helmet. With the help of embedded sensors, CHECKLIGHT measures head impact during competition.

Golfers are beginning to use sensor-enabled golf clubs that can track their swing and provide relevant analytics. Zepp offers a sensor-impregnated attachment for your baseball bat. By digitizing your swing you are able to overlay it on the swing data of a pro. Skiers can now wear ski goggles that track a large number of metrics in real time.

The startup 94fifty showed a basketball at CES that has several sensors embedded in it. Together they measure elements of your shot such arc, speed, rotation, and height. But more than providing raw data, they provide curated feedback through their app. Did your shot have too little arc? The app—relying on the digitized sensor data—can provide input and suggestions to fix your shot in realtime.

Seeing today's intersection between digital data and athletics, we can envision a time when concerns about concussions and other sports-related ailments might become negligible. But we aren't there yet. Using digital data to enhance athletic performance is new territory. Sensors data could open entirely new veins of inquiry and ultimately change some sports as we know them today.

MANAGING OUR DIGITAL SELVES

With digital storage quickly becoming limitless, consumers are growing tired of managing the massive amounts of digital data they are inundated with on a daily basis. Plus, individuals today are quite aware that the Internet never forgets. We want to control our digital lives just as carefully as we control our analog lives. We want to control what we share and whom we share it with—but we understand that the pervasive properties of digital may not leave us that choice. Today, a growing number of apps and programs designed to let you be you—in all of your unfiltered, unshaven, unmannered glory—but still have a say in your digital destiny are quickly attracting followers and skyrocketing pricing valuations.

Privacy and anonymity are the currency for many of these new apps and services, which let you share personal thoughts and photos. Snapchat was one of the early entrants to a quickly widening field. Self-destructing and encrypted message app Wickr, for example, raised $9 million in a Series A round of financing. Whisper, an app that lets you post messages and receive replies anonymously, raised another $30 million at an astounding $200 million valuation. The latest anonymity app to make headlines is Secret, which lets users post updates under semi-anonymity. Within a week of its launch, Secret raised $8.6 million in new funding.

After a decade of over-sharing online, consumers are starting to appreciate the value of communicating in an anonymous and ephemeral way, and they are seeking out privacy-minded apps and services.

Clearly defined privacy policies are becoming a badge of honor, so much so that companies are building and betting their entire existence on well-defined privacy policies. Jan Koum, the founder of WhatsApp, turned to the company's blog to defend its acquisition by Facebook. "Respect for your privacy is coded into our DNA, and we built WhatsApp around the goal of knowing as little about you as possible," Koum wrote. "We don't know your likes, what you search for on the internet or collect your GPS location. None of that data has ever been collected and stored by WhatsApp."

As more anonymity platforms flood the market—WUT, Social Number, YikYak, Shrtwv, Banter, Blink, Backchat, ask.fm—the battle is on to live up to the promise of security and privacy. And as these services show signs of going mainstream and becoming less secure, then potentially more secure tools get big boosts. The market for anonymity is growing steadily, as consumers look for outlets for more personal expression in our pervasive digital environment. But these large increases in user numbers are an indication of the widespread desire to control our individual digital destinies.

Thanks to these apps, our digital existence is narrowing to more closely mimic social interactions in our real lives. If Facebook is a sprawling suburb filled with co-workers, second cousins, and your kid's soccer coach, then the anonymity app is the tiny city apartment you shared with four roommates in your twenties. These are the people who saw you shuffle around in your pajamas, knew your favorite bands, and drank your milk. We feel comfortable sending these types of friends unflattering five-second snaps that we'd never use as our Facebook profile photo—they love us at our worst. These new semi-secret communities are different from "social media" as we know it. We're entering an era in social media that rewards disclosure and sincerity over perfection. The question is, will this lead to a more authentic Web—one in which we control the distribution of our posts and therefore no longer feel the need to censor ourselves?

These apps provide digital communication channels that let us be ourselves with the people who matter most to us—especially those we don't get to see every day—without the embarrassment of over-sharing. Our enthusiasm for these apps isn't about wanting more privacy. In many instances our motivation is just the opposite—we want to share more. (Some might say we want to share it all.) But we want to decide how, when and with whom we share.

These digital communication channels haven't been created to circumvent accountability. They are giving users greater control over privacy by enabling control over identity or distribution. They are placing greater control in our hands when it comes to managing our online identity. And in the process, they are spawning entirely new ways for us to communicate and interact.

THE MIND MELD: WHEN DIGITAL DATA AND IDENTITIES MERGE

Recently I was driving in my Ford Explorer and began to use the voice-activated navigation system, which responds to voice prompts with what are essentially questions, as it determines where I want to go. My six-year-old son Gavin was riding along and asked, "Is that Siri, Dad?" I laughed as I replied that it wasn't, but his next question was even more interesting: "Well, what's her name?" You'll note he didn't ask what *IT* was called. He used the pronoun "her." Immediately I recognized that my son was growing up in an environment where communication is increasingly digitized in a way that appears very human. We have traveled the length of the continuum of computer-human communication in a very short time. I reflected on the movie *Her* and realized that what seems strange and maybe even slightly creepy to us seems perfectly normal to our children.

The digital destiny before us is about so much more than just digital data. It isn't about new devices, or some story on the back page of some newspaper about how something that was once analog is now digital. Our digital destiny has massive cultural implications that will produce huge shifts in everything: How we talk. Live. Work. Communicate.

Kids are growing up in this environment, and it will be the new normal. Many of the concerns occupying us today are simply part of the usual chaos that occurs when massive amounts of data get unleashed. In time, order will descend, alleviating our concerns as we learn to live in a world where the division between physical life and our digital existence has dissolved. We might also hope that, with this expansion of digitized information in our lives, many of the unsavory behaviors we associate with the Internet—cyberbullying, loneliness, relationship drama—will gradually recede. Though as we've seen in countless other examples, the chaos will eventually break out elsewhere as the cycle begins anew.

CHAPTER 13

Economics and Business in a Digital Age

"Everybody believes in innovation until they see it. Then they think, 'Oh, no; that'll never work. It's too different.'"

—Nolan Bushnell

I n August 2014 a curious thing happened at the intersection of economics and politics. The ride-sharing company Uber announced the hiring of David Plouffe as senior vice president of policy and strategy. Plouffe, politicos may recall, managed both of President Obama's successful campaigns for the White House. If the 2008 campaign made Plouffe one of the top political strategists in the country, then the 2012 campaign sealed his legacy as an early pioneer of data-driven politics.

So why in the world would an app-based ride-sharing company hire a political consultant to lead its policy and strategy? Even more curious is why Plouffe would join a company that the *Republican*

National Committee had recently heralded in an online petition of support.

As the RNC petition stated, "across the country, taxi unions and liberal government bureaucrats are setting up roadblocks, issuing strangling regulations and implementing unnecessary red tape to block Uber from doing business in their cities. We must stand up for our free market principles, entrepreneurial spirit and economic freedom."[1] The petition was in direct response to the efforts of many cities to either force Uber out of the market or else make it abide by traditional taxi regulations. The fact that nearly all of the cities are managed by a Democratic mayor and a Democratic council just add to the curiosity of Plouffe's hiring.

One can't guess Plouffe's full motivations. But Uber's are more easily seen. Precisely because the company is aggravating entrenched political interests it has turned to a seasoned political operative to help Uber navigate the rough times ahead. That Plouffe also knows his way around digital data and how to build upon a foundation of loyal supporters surely made his hiring a unique, irresistible opportunity.

In its short five years of existence, Uber has catapulted to the top of the burgeoning data-driven economy. By connecting passengers with drivers, Uber and other ride-sharing companies such as Lyft and SideCar tapped into a consumer desire for a simpler, more convenient taxi service. But as so many other tech companies have had to learn the hard way, enormous success has brought Uber into direct confrontation with a legacy industry and its political allies.

Across the country, taxi companies have lobbied their political networks to make life for Uber as difficult as possible—and sometimes that means making life for regular people as difficult as possible too. In Washington, D.C., for instance, a coalition of taxi companies staged a protest in the middle of Pennsylvania Avenue—blocking traffic and honking horns to signal their displeasure with Uber's and Lyft's existence.

Despite the RNC's petition, the battle lines aren't as neatly drawn along party lines. Virginia governor Terry McAuliffe, a Democrat, lifting a temporary cease-and-desist order from a state agency, worked with Uber and Lyft to reach a compromise. Likewise, the Council of the District of Columbia, where Democrats hold eleven of thirteen seats (the other two are held by Independents), has generally welcomed ride-sharing companies without much red tape. The D.C. Taxicab Commission is another matter.[2]

Still, when you consider the forces aligned against Uber—a coalition that includes taxi companies, unions, regulators, and politicians—it's remarkable the company and its competitors have survived at all. Remarkable, but hardly unprecedented. With some notable exceptions, Americans, our politicians, and our courts almost always side with technology and the progress it provides. It's just one reason American tech companies are the pride of the world, their products valued by people in every country on earth.

The taxi companies and their allies are fighting a rearguard action. One suspects they know it, too. It's in their interest to oppose the forces of progress such as Uber for as long as possible, if only to forestall the revenue declines that accompany disruption. The technology available today makes the old ways of the taxi industry nearly obsolete—just as the personal computer made the typewriter obsolete. This is not to say that the industry can't survive if it updates its business model and equipment. But the days of being a protected industry without competitive pressure are over.

If by chance the industry manages to destroy Uber and its peers (as unlikely as that might be), another Uber is just around the corner, and another one after that. Having been exposed to Uber, consumers won't put up with the inefficiencies of the old taxicab ever again. Nor will tech entrepreneurs simply stand aside when a proven alternative to taxicabs goes unutilized. The genie is out of the bottle.

I should add that I have no desire to dance on the grave of the taxi industry—or any industry that will get bulldozed by digital data. As an economist, I fully appreciate that an industry is more than numbers on a page. Industries are jobs. Industries support families. But as an economist, I also appreciate that attempts to block technological progress always lead to worse economic outcomes. Nevertheless, telling a cab driver that he's going to lose his job so that future families may thrive is cold comfort. But it's the economic reality.

Our digital destiny is about more than the prospect of a better life. It's about more than privacy concerns or driverless cars. It's about facing this economic reality, the certainty that to create something new and better, something old and worse must be disrupted. It's about the workers and the families who will be displaced by the upheaval we're already experiencing. As history has shown, however, it's also a reality that promises more wealth, more freedom, and more prosperity than what has come before.

THE GREAT DEBATE

The acceleration of broad digitization and its implications have engulfed some of today's leading thinkers in a great debate regarding the impact digitization is having on the broader economy. In one corner you have skeptics such as Jaron Lanier, a computer scientist and writer who coined the phrase "virtual reality":

At the height of its power, the photography company Kodak employed more than 140,000 people and was worth $28 billion. They even invented the first digital camera. But today Kodak is bankrupt, and the new face of digital photography has become Instagram. When Instagram was sold to Facebook for a billion dollars in 2012, it employed only thirteen people. Where did all those jobs disappear to? And what happened to the wealth that those middle-class jobs created?... Instagram isn't worth a billion dollars because those thirteen employees are extraordinary. Instead, its value comes from the millions of users who contribute to the network without being paid for it. Networks need a great number of people to participate in them to generate significant value. But when they have them, only a small number of people get paid. That has the net effect of centralizing wealth and limiting overall economic growth. Instead of enlarging our overall economy by creating more value that is on the books, the rise of digital networking is enriching a relative few while moving the value created by the many off the books.[3]

It's abundantly clear that Lanier is skeptical of digital's ability to create lasting and widely distributed wealth. The growing income inequality in America and much of the Western world only would tend to support this viewpoint, and Lanier isn't alone in those views. It is a perspective also argued by Andrew Keen in his most recent book, *The Internet Is Not the Answer*. And software developer Karl Fogel, author of the influential *Producing Open Source Software*, believes, "we're going to have to come to grips with a long-term employment crisis and the fact that—strictly from an economic point of view, not a moral point of view—there are more and more 'surplus humans.'"[4]

But not all share this apocalyptic view. On the other side of the debate are folks like Hal Varian, professor of information sciences, business, and economics at the University of California at Berkeley and chief economist at Google: "How unhappy are you that your dishwasher has replaced washing dishes by hand, your washing machine has displaced washing clothes by hand or your vacuum cleaner has replaced hand cleaning? My guess is this 'job displacement' has been very welcome, as will the 'job displacement' that will occur over the next 10 years. This is a good thing. Everyone wants more jobs and less work."[5]

Or Tyler Cowen, Holbert C. Harris professor of economics at George Mason University, who has provided perhaps a more balanced view: "The law of comparative advantage has not been repealed. Machines take away some jobs and create others, while producing more output overall...yes robots may lower employment, although the catchphrase 'robots are destroying jobs' is misleading rather than illuminating."[6]

The thing about technology is that it just keeps getting better. The move toward digital both makes this improvement more encompassing and accelerates its impact. Digital is hitting more things than ever before, and broader integration of sensors pushes this trend further. Technologically-induced change continues its march forward and picks up speed when it hits digital. Everyone sees this; we just disagree on how exactly the change is going to march forward once it hits us.

That it will hit us is undisputed; but it's going to hit different people in different ways. There's an old joke among economists that a recession is when your neighbor is unemployed and a depression is when you are unemployed. By that logic, as digital data impacts each of us differently, some will be experiencing a recession and others a depression—while still others are thriving. Some jobs are in greater jeopardy than others, while newly created jobs are just now entering the picture as well; some industries are facing an existential threat, while others are just realizing their potential.

Indeed, without much effort we could make a list of industries that have already heard the indifferent bulldozer of digital data knocking on their gates. We could add to this list certain jobs, common in many industries, that will likely disappear over the next decade or so. We can make a fairly good guess at which industries stand to gain the most as well as which jobs will be in the most demand.

What we can't predict very well is what industries and jobs will be around beyond our ten-year window. You may have noticed that throughout this book I've avoided saying that this device or that product will be the device or product we will use in the future. I have mentioned several specific companies whose products reveal future trends and technology's potential, but I can't perfectly predict who or what will be the Steve Jobs or Google of the future.

But I can safely predict one thing.

As Hal Varian said in 2009, the sexy job in the next ten years will be statisticians. Carrying this logic further, the role of data scientists will redefine nearly every job—sexy or not—in the next decade. We can see this happening today in diverse industries. Consider the newspaper industry, one of the most disrupted by our transition to digital. The *Washington Post* recently listed job openings for "data journalists" to join a team of writers and editors in a new venture. (Jeff Bezos' influence is having the desired effect.) Not only is reporting on the news being disrupted by our digital march, it is also being redefined by data. A similar trend is unfolding in every industry.

Indeed, part of the explanation for the prolonged high unemployment level of recent years—and I stress the word *part*—is that the market crash of 2008 and subsequent recession of 2009 forced the economy to slim down and redefine jobs with an eye towards the inevitability of digital. Under good economic conditions these jobs were temporarily safe, because employers weren't forced to make hard choices. But once economic conditions turned sour, forcing employers' hands, those jobs were lost—in many cases permanently. The industry or company learned to live without them; processes were modernized with digital components, rendering the old job superfluous.

As Erik Brynjolfsson and Andrew McAfee, two economists from MIT and authors of *The Second Machine Age*, put it, "rapid and accelerating digitization is likely to bring economic rather than environmental disruption, stemming from the fact that as computers get more powerful, companies have less need for some kinds of workers."[7]

This has left millions of displaced workers with nowhere to go. Even if they were trained in the new digital jobs of the future, those jobs aren't in abundance yet—certainly not at the level needed to make up for the great displacement of 2009. In time, I believe, those changes will occur, creating new jobs for a new generation of workers and redefining nearly every existing job through the lens of digitized data.

So in the matter of the Great Debate before us, I side with Hal Varian and Tyler Cowen: The law of comparative advantage has not been repealed. As a country, we may have underestimated the severity and longevity of the displacement, but we shall emerge, in time, and when we do, we will find an economic landscape very different from the one we left in 2008.

EMPLOYMENT AND DIGITAL AUGMENTATION

Here's the thing. We are replacing some types of labor with capital—just as the printing press did. This trend has been observable for thousands of years. Digitization is accelerating and intensifying that trend. If the jobs we are taking away have a low digital component and the jobs we are adding have a high digital component, then we have a lot of work to do to train tomorrow's workforce. We are digitizing everything, which means that all jobs will be touched by digital in some way. It's the degree of digital's influence that is the question.

Digital technology will first replace jobs that involve repetitive, easily defined tasks. Digital technology will also significantly impact jobs with a heavy information component. At the same time, the prices of unique, non-tradable skills will increase. Much of the discussion around jobs in a technological world revolves around the simplistic idea that "tech takes jobs." But digitization of information is less about replacement and more about augmentation. Tomorrow's jobs will all be infused with digital data, from the farmer relying on data captured by digital sensors to school teachers redefining what material they cover and how they teach based on digital measures of student retention and understanding.

It's helpful to look at examples across several industries. The number in parentheses represents the percent of total workers in the United States currently employed in the respective industry:

Retail and Wholesale Trade (14 percent): Sensors are being used today across the entire supply chain of a product—from the moment it exits a factory until it is bought off a store shelf. The newly available digital data is utilized in predictive analytics applications to help optimize inventory across the supply chain to ensure inventory levels are appropriate for a given product. Technologies like Apple's iBeacon are enabling retailers to 'sense' nearby smartphones and provide additional information about available products. Companies like Procter & Gamble and Unilever have begun utilizing eye-tracking technologies to measure and test packaging design concepts with consumers. Lowe's has begun testing fully automated robots that utilize sensors to navigate the aisles of their hardware stores. The OSHbot relies on natural-language-processing technology to answer customer questions. Using embedded imaging sensors, the OSHbot can even 3D scan an object and help customers identify what they are looking for.

Transportation and Warehousing (3 percent): FedEx and UPS are well known for sophisticated tracking systems that enable you to see the location of your package at any moment. But they also track their trucks, using more than two hundred sensors on every truck. At the time of the rollout of this technology, Donna Longino, a UPS spokesperson, said, "Telematics isn't new to UPS. We've been using telematics for more than 20 years to improve the efficiency and safety of our tractor-trailer fleet. What's new is the proprietary information and sophisticated algorithms we developed to analyze the rich stream of data captured by more than 200 sensors on our delivery trucks." It isn't just collecting the data—it's making the data actionable.

UPS is not only monitoring the location of their trucks, but they've also installed sensors on things like the brakes and the engine box that will help them make adjustments and improvements around idle time and routing efficiency. For example, this data has helped them create routes that avoid left turns to minimize idle time, increase productivity, efficiency, and lower cost.

Shipping drivers, therefore, may log fewer hours—as digital data improves their productivity. But they still have their jobs—for now, at least. We can easily envision a future in which driverless shipping trucks replace human drivers. We can also envision a day when drones (such as the ones that Amazon is experimenting with) replace trucks altogether. The benefits of modernized shipping—whether more efficient routing, autonomous vehicle utilization, or drone deployment—are impossible without the wide deployment of sensors.

Leisure and Hospitality (9 percent): Starwood property Cupertino's Aloft Hotel recently deployed a high-tech butler in the form of a three-foot-tall robot called A.L.O. Botlr. Just press its seven-inch tablet screen and the "butler," designed by Savioke, can call for the elevator, roll up to your door with whatever you requested, and call your room phone to let you know it has arrived. Botlr uses sensors to digitize and navigate its physical surroundings. Hotel 1000 in Seattle has begun utilizing silent infrared doorbells with heat sensors that detect the presence of individuals inside a hotel room and enable hotel staff to determine if the room is occupied before servicing the room.

Manufacturing (8 percent): Manufacturing has long been home to the industrial applications of sensors. However, to date most sensors have been deployed in fixed automation. In 2013, the company Rethink Robots released Baxter, not just another ordinary run-of-the-mill robot who can perform a variety of human functions. Rather, Baxter can be programmed to do a number of tasks its developers have not yet even conceived—allowing it to be used in high-mix environments and redeployed to different tasks as need arises. Rethink Robots envisions an "app-like" market for robots like Baxter, in which developers create programs designed specifically for certain industries, which can then be uploaded to Baxter.[8] Baxter relies on a number of embedded sensors including cameras, force sensors, sonar, and rangefinder, which allow it to adjust to the subtle changes of a dynamic environment just as you or I would adjust.

Financial Services (5 percent): Sensors and digitized information have long impacted financial services. Some of the earliest customization and personalization on the Web was creating tailored lists of stock tickers. Today the digital aspects of our smartphones have turned these pocketable devices into bona fide banks. Not only can we perform rudimentary activities like checking our account balances, but we can also utilize the image sensors on our smartphones to digitize and deposit physical checks. Insurance companies are increasingly relying on digitized data to price insurance policies. Progressive's Snapshot program is a usage-based insurance approach that prices policies based upon data transmitted to the company from the vehicle's OBD-II port. Rather than solely relying on a coarse measure like FICO scores to measure creditworthiness, financial service startups like Affirm are beginning to measure creditworthiness using all available digital information about you—even, for example, data like your Facebook profile.

Government Services (8 percent): Yardarm Technologies has developed a sensor array currently being tested by several local police departments. The sensor unit attaches to a police officer's handgun and through a variety of sensors like accelerometers and gyroscopes digitizes a wide range of information including when the gun is unholstered and if it is fired. This information is then communicated in real time to the police department, which can monitor when and where the gun of every officer is being used.

Educational Services (9 percent): Universities and colleges are another industry where digital data stands to make a huge impact. Already the rise of Massive Online Open Courses (MOOCs) has foreshadowed a massive transformation in the way in which students will learn. Currently, education, particularly at the higher levels, is limited by "shelf space"—in that only a certain number of people can fit into a classroom at any given time. But with the Internet, "shelf space" is eliminated in education just as it's eliminated in retail. With online courses, universities no longer have to worry about class size. Moreover, the availability of an online lesson frees the student from fitting his life around a university's schedule; now he can take the online course whenever he wants.

Obviously, there are drawbacks to the MOOC system. It becomes nearly impossible to get personal attention from the professor when class sizes number in the thousands. Yet some schools, such as Western Governors University based in Denver, have struck a balance. WGU classes are taught entirely online in real time, and the professors also made time to meet with students via video apps. By all accounts, it is a well-received compromise, as WGU and other online institutions have exploded in popularity in recent years.

Digital formats such as MOOCs and online universities work especially well for a type of learning that diverges from the standard university model. Not everyone desires or needs a higher education experience that includes standard subjects such as the humanities and science. Technical and trade schools don't necessarily need the one-on-one professor-student dynamic that your typical bachelor's degree requires.

The digitization of information is flipping the historical understanding of the classroom on its head. The Flipped Classroom movement is an approach that makes the traditional classroom lectures available via online video so students can watch them at home while taking the assigned work and problem solving into the classroom. By taking what we've always considered "homework" and doing it in the classroom, students will have access to more interaction and personalized guidance from teachers. Emerging services like Khan

Academy allow teachers to digitally monitor the progress of their students. Digitization is changing a process of learning that has been in place for thousands of years. The entire education system is changing right before our eyes.

BUSINESS MODELS IN THE NEXT DIGITAL ERA

Clayton Christensen in his seminal book *The Innovator's Dilemma: When New Technologies Cause Great Firms to Fail* explains that market-dominating companies, often large and entrenched, are prone to disruption by upstarts precisely because incumbents are lulled by the comfort of their entrenched positions and fail to sufficiently heed encroaching threats from new market entrants. Instead they focus on protecting their market share from existing, similar competitors. While seeking to serve their core customers, market leaders slowly cede segments of their market until there are no pieces left to relinquish. Upstarts gain a foothold in these well-served markets with solutions that are simpler and less costly, and (at least initially) not as good—the key to why incumbent players fail to immediately worry about the newcomers. Rather than compete in low-margin segments of their industry, they focus on higher-margin segments where greater profitability can be had. Eventually, what begins as a small segment of their market grows larger as the upstart itself matures and gains a greater and broader industry hold. This slowly moves the upstart into segments of higher margins until it is actually the dominant player.[9] John Hagel, co-leader of the Center of the Edge at Deloitte, put it best when he recalled Joseph Schumpeter's idea of "creative destruction" and explained that "markets are a powerful engine for 'creative destruction'—they invite competitors with a better idea or a better approach to come in and challenge incumbents."[10]

In the last twenty years, a tremendous amount of attention has been given to understanding how disruption occurs, but we benefit most from a contextual understanding of disruption. The original intention behind disruption analysis was to shift attitudes away from avoidance and reluctant response, but what actually happened was that disruption became an infallible protagonist, a disruptor itself. In the narrative we've now created, Drake Bennett of *BusinessWeek* explains, "the upstarts are the heroes. Their eventual victory over the established order is foreordained, and they are the force that moves society—or at least technology—forward, disruption by disruption. Starting a company holds the potential to be not only lucrative, but also revolutionary."[11]

Many indications suggest disruption is accelerating. For example, Hagel points to the topple rate, a measure of how fast companies lose their lead positions in their respective industries. The topple rate has increased by almost 40 percent since 1965, and the tenure of companies on the S&P 500 (a rough list of the largest five hundred publicly traded companies in the United States) has fallen from seventy-five years in 1937 to eighteen years today.

DISRUPTION IN THE NEW DIGITAL ERA

We are entering a new digital era with profound implications for the future of disruption. To recap, the first digital era began when analog devices were replaced by their superior digital equivalents, but this next digital era is driven by the broader digitization of our physical space. In the second digital era, not only will we digitize more of the objects around us, but we will systematically digitize the information the objects gather. It is this digitization of data that fully opens the floodgates of disruption. As Hagel explains, the difference between the major technology disruptions of the past (the steam engine, electricity, and telephony) and today's disruptions lies in digital technology's demonstration of "sustained exponential improvement in price/performance over an extended period of time and continuing into the foreseeable future.... there's no stabilization in the core technology components of computing, storage and bandwidth."[12] The stunning acceleration of disruption today is in part explained by our entrance into this new digital era.

THE REALITY OF STARTUPS

While instinct tells us to credit the leadership of successful upstarts for being disruptive, some of the credit actually should fall to the environment in which the upstart finds its footing. Disruption is, after all, a process; and upstarts owe some of their success to their position within the process. In the beginning, the inherently lower cost structure of the upstart allowed it the latitude to experiment. While many of these experiments go nowhere, some experience tremendous success. Another key to disruption is the startup's ability to build its industrial structure and operating procedures on a foundation of current technologies, so that it can produce services and products on the cutting edge of value and application in the current business environment. Startups often appear disruptive vis-à-vis giant incumbents, but we can say more accurately that they are able to differentiate themselves partly because of their structure at the time of their genesis. According to Hagel, "disruption turns the assets of incumbents into potentially life-threatening liabilities," and upstarts are not saddled with these liabilities—at first.[13]

THE (BASIC) ECONOMICS OF DISRUPTION

In competitive markets, disruption will always drive an industry's net growth, as a function of simple economics. New companies typically have lower cost structure born of the features of brand new labor force or the deployment of new, more productive capital. The entrant with a lower cost structure will drive down the marginal cost in the industry—which, in a competitive market, means a shifting out of the supply curve. In a competitive marketplace, demand is equal to price so that, even without changes in preferences, the industry's new cost structure means increased demand for a greater quantity, and as a result the total market grows. While we often recognize some intrinsic acumen on the part of the disruptors, a significant portion of their success is the result of the basic economic principles at play.

CHARACTERISTICS OF DISRUPTION IN THE NEW DIGITAL ERA

If measures of disruption are accurate and disruption is in fact accelerating, it is worth considering the face of disruption in the digital era before us. Factors helping to drive disruptions in the new digital age include the ability to scale innovations, the network effects inherent in digital environments, compressing diffusion cycles, a shift toward services, the development of new platforms, and the characteristics of those platforms.

Scale: Scaling businesses has always been a key element of growth, but the character, cost, linear nature, and magnitude of scaling businesses changes significantly in the new digital era. Switching costs are lower in a digital marketplace. In an analog world, consumers face real costs when choosing to find and travel to a new store or engage a new service provider. But in a digital marketplace, both search costs and switching costs are lower, meaning consumers can more readily find and switch to new businesses without major disturbance to themselves. Lower switching costs enable businesses to scale (and fail) quickly. The lack of real-world tangibility also accelerates scale. Moreover, digital businesses can scale their core business without scaling tangible assets in tandem, meaning digital businesses face increasing returns to scale. In other words, they can grow revenue faster than the costs required to capture the new incremental revenue.

Digital has also changed the magnitude associated with scaling a business. In an analog setting, selling the first fifty thousand units might be the threshold that suggests the marketplace is viable. The scaling of Internet businesses today is gauged at a completely different tier. Today we might say anyone can get his app downloaded 1 million times; it's the next 10 million times that matter. One reason behind this shift in scale is that digital enables a hyper-scaling that is fueled by not only rapid scaling but also wide scaling.

Network Effects: In economics, a network effect is the effect users of a good or service have on the value of that good or service to all other users and potential users. Network effects have always been a powerful force in technology and help explain how the value of devices like telephones or fax machines increase (and decrease) over time. For example, the value of the first fax machine was close to zero without a second fax machine from which to send and receive documents. But as the number of fax machines began to steadily increase, owners of fax machines had more and more people from which to send and receive documents. The value of the network increased as more users joined the network. The same is true as users of fax machines began to leave the network. Positive network effects are almost always a defining attribute of disruptive businesses in the second digital era. Today, platforms like Instagram, Pinterest, and Snapchat all benefit from positive network effects.

Compressing Diffusion Cycles: One of the properties of the new digital era is the ability to quickly determine if there is a viable marketplace for a new digital offering. Historically, companies could innovate slowly around core businesses to both create and capture value. Innovation would take place linearly with time. It could be slow and methodical, but that is no longer the case. Upstarts are now creating value around existing business from all sides. Digital compresses diffusion cycles, which, in turn, compresses adoption cycles, meaning startups are able to create entire businesses as well as entire industries near-instantaneously. In the new digital era, we are moving from a world of incremental innovation to one in which value is created and organized quickly. In the analog world, if one wanted to enter a given market—even a service market—one had to organize capital. Expansion would require capital because businesses didn't generally have increasing return to scale. In other words, they had to grow costs in order to grow revenue. But in the new digital era businesses generally enjoy increasing returns to scale, so they can expand quickly. At the same time, as I pointed out earlier, prospective adopters of the service face lower search costs and lower switching costs. These forces combine to compress diffusion of the new business and ultimately adoption of the service.

A Shift to Services: Over the last fifty years, we've had a broad shift towards services. In the 1960s, U.S. consumers spent about 45 percent of total spending on services. Today, that same figure has increased to 66 percent,[14] and the real figure is likely higher when you take into account the higher number of devices that rely on accompanying services to maintain their value to end users. The products and devices launched in today's digital environment look increasingly more like services. Furthermore, the entire design cycle and subsequent product cycle of digital products are undergoing tremendous change. Because software is a large component of a device today, manufacturers are able to update, upgrade, and change the functionality of a product after the point of purchase.

Historically, if consumers wanted to upgrade their devices to the newest available applications, they had to buy the newest hardware. Now, because hardware is a smaller portion of the overall experience, manufacturers are increasingly able to push upgrades to devices. In some cases, software upgrades are even changing what was the original use for the device.

Multisided Platforms: The theory of disruption was the lead theory on change over the last decade and a half. But the economics of multisided platforms is set to replace disruptive theory as the key theory on business organization and industry dynamics over the decade ahead of us.

Multisided platform businesses are best thought of as marketplaces that allow participants to have direct interaction with each other. The scaling and network effects of digital environments create an atmosphere well suited for multisided platforms, or what economists call two-sided marketplaces. These marketplaces contain two distinct groups, and individuals in each group benefit from interactions with individuals on the other side. Generally the two distinct groups in these marketplaces provide each other with network benefits. Examples include temporary staffing agencies (workers and employers), search engines (advertisers and users), credit card networks (merchants and cardholders), and even shopping malls (shoppers and merchants). Multisided platforms enable direct interaction between individuals from the two distinct groups, and that interaction improves opportunities to trade in the marketplace through network effects. The number of successful digital multisided platforms seems to be growing daily. Examples include Uber, Airbnb, Square, Craigslist, and even sites such as Tinder or Wechat.

Examples of Digital Multisided Platform Businesses	What They Do
Airbnb	Helps travelers rent homes, apartments, or rooms directly from owners.
Bla Bla Car	Facilitates car sharing (Bla Bla Car is Europe's largest car share).
Boatbound	Allows for boat rentals directly from the owners.
Cargomatic	Connects local shippers with carrier companies who have extra space in their trucks.
Circle Up	Links accredited investors with consumer product and retail companies.
Deliv	Crowdsources same-day delivery service for large national multichannel retailers.
Elance	Online platform for connecting freelancers and businesses.
Hailo	Connects taxis and passengers.
Instacart	Allows for same-day delivery for grocery and home essentials from a variety of local stores.
Lending Club	Facilitates direct peer-to-peer lending via an online network.
Lyft	Matches passengers looking for rides with drivers offering rides.
Our Crowd	Crowdfunds startups with venture capital through an equity-based platform built exclusively for a select group of accredited investors.

Examples of Digital Multisided Platform Businesses	What They Do
Pivotdesk	Matches companies together to share office space.
Postmates	Delivers anything that you order via a phone app to your door in under an hour.
Prosper	Enables individuals to either invest in personal loans or request to borrow money.
Relay Ride	Allows private car-owners to rent out their vehicles.
Sidecar	Matches riders with drivers.
Soundcloud	Enables music creators to upload, record, promote, and share directly through an audio platform.
Skill Share	Hosts an online learning community.
Storefront	Provides short-term retail spaces for rent to companies for pop-up stores.
Traity	Verifies the online profiles and reputations of individuals.
Uber	Connects passengers with drivers of vehicles for hire and ridesharing services.
Wattpad	Allows users in a writing community to post articles, stories, and poems.
Yerdle	Provides daily list of things people are giving away.
Zopa	Matches individual borrowers and lenders in the U.K.'s largest peer-to-peer lending service.

Because of the network effects of multisided markets, monopolies tend to form wherever platform businesses thrive. Coordination among individuals will be easier because we'll all gravitate to the same platform, but the risks of abuse are also higher.

BEYOND THE GREAT DISRUPTION

The literature since Christensen's seminal work has constructed and explored a framework for identifying and explaining patterns of disruption—for both the disruptor and the disrupted. Changing business environments present new patterns and paths of disruption, and the next decade will bring with it not only new challenges but also new pathways to success. As Hagel has written recently, incumbent players must "find ways to expand the horizons of their leadership team beyond the next quarter or next year and to challenge on a sustained basis the key assumptions, often unstated, that they bring to the table regarding what is required for business success."[15]

The new digital era brings with it an environment extremely conducive for disruption and change. Multisided platforms are developing and disrupting incumbent business models. In turn, more seeds of disruption are being planted today that will ultimately exert pressure on multisided businesses. The attributes and characteristics of digital suggest that disruption, in all its varied forms, will only accelerate.

A theme we've explored repeatedly throughout this book is the idea of order from chaos. With economic upheaval and the struggle of entrenched industries against tech startups, we can say that we are in the middle of a great deal of chaos. We can also say with a great deal of certainty that order will reemerge, giving both shape and meaning to the chaos around us. But what we need to understand is that, just as digital accelerates and intensifies almost everything it touches, so too does it accelerate and intensify the economic cycle. In other words, the periods of relative economic stability that the world has enjoyed in the post-war era—when technology, business models, and consumer behavior were static for a time—is likely a relic of the past. In our digital future, Schumpeter's "creative destruction" will be continual, as digital data relentlessly pushes us beyond our expectations. The whole notion of "legacy industries" will be lost to the ages. Every industry, every company, indeed every job, will be in a constant state of flux and renewal. Swirling eddies of chaos and order will surround us, the dynamic cycle of creation and destruction simply a fact of life. It is a frightening prospect precisely because it will be an entirely new experience for most of us.

But if our experience with digital so far has taught us anything, it's that the younger generations are quite capable of living, working, and thriving in this new world. They will know nothing else. And the comfort we take from their ability to compete and succeed in the whirlpools of chaos and order will help us remember that the world we leave them will be freer, richer, and more personal than anything we have known.

CHAPTER 14

The Parallel Evolution of Digital Data, Law, and Public Policy

"We can only see a short distance ahead, but we can see plenty there that needs to be done."

—Alan Turing

I n the early days of digitization the online world was viewed as a separate domain—an alternative reality. We *went* online. Today however, as we have seen, we *are* online. Period. Once parallel existences have now converged. When our online and offline lives were clearly independent it was possible to maintain dual identities—not so much in the traditional sense, but at the most basic level of our existence in these two diametrically different worlds. What we did online was broadly different from what we did in the rest of our life. But over time digitization reached more segments of the economy. Before we knew it, we were doing things such as shopping or banking in both the physical and digital worlds. Furthermore, the divide between what were once different identities became increasingly blurred in a world where elements of our digitized identity influence our physical identity and vice versa.

Our identity has become inextricably interwoven with digital data. With Twitter identifying gender not through physical traits or self-selection but through emergent user behavior patterns, data is defining us as we define data. We are the data and the data is us. Our identities are becoming equally defined by what is happening online and offline, and this has broad implications for our to relationships, sense of self, and self-identification—but also for things like our sense of citizenship and rights and even the definition of democracy.

Digital data is piercing more deeply into every segment of our everyday lives, both private and public. Arguably, digitization has even changed the definition of private life, making it possible for others to see, through digital platforms such as Facebook where one can broadcast anything quickly and easily, facets of our lives that were not shared before. As our partitioned identities fuse, a similar knitting together will take place in public policy. What historically have been two relatively easily discernable and detached spheres will merge.

EVOLUTION OF THE LAW

On August 22, 2009, David Leon Riley was arrested in San Diego after a traffic stop revealed loaded firearms in his car. Upon searching Riley, the police discovered his cell phone. Using evidence such as videos and pictures contained in the phone, the police would eventually charge Riley in a shooting unrelated to his traffic stop. Riley's case ended up in the Supreme Court. Sometimes it's such random encounters that prompt major legal decisions and create new frameworks.

To fully understand the implications of Riley's case we need to take a couple of steps back. The police were fully within their constitutional rights in searching Riley's person. In *Chimel v. California* (1969), the Supreme Court ruled that the police may search the body of an arrested suspect and the immediate vicinity without a warrant. This was for the protection of the arresting officer. But in 1969 arrested suspects did not carry with them devices that contained large amounts of personal—and potentially incriminating—data. While *Chimel* still governs the actions of arresting officers to this day, Riley's case highlighted the need for a clarification.

It would not be the first case that the Supreme Court had heard related to digital data and personal privacy. In the 2012 case *United States v. Jones*, the Supreme Court ruled that attaching a GPS device to a car and monitoring its movements constitutes a search under the Fourth Amendment. *Jones* was narrowly decided, but it did augur that cases involving personal data—whether tracking a person's movements via GPS or whatever information is on a phone—were about to become much more common.

It took nearly a decade for the *Jones* case to wind its way to the Supreme Court—a decade in which our mobile phones became increasingly sleeker and smarter. And as mobile technology evolved, so too did consumer demand. Today, 90 percent of Americans own mobile phones and more than half own smartphones; many of us can't remember a time when we weren't tethered to our devices.

In 2014, the court issued a single, unanimous opinion on *Riley v. California* and *United States v. Wurie*, finding that police must obtain a warrant in order to search digital information on a cell phone seized from an individual who has been arrested. In delivering the court's opinion, Chief Justice John G. Roberts Jr. noted that, for many Americans, today's smartphones hold the "privacies of life," adding that the American Revolution itself was predicated, in part, on opposition to illegal searches.

Chief Justice Roberts also highlighted the difficulty of setting legal precedents that won't be outstripped by technological innovation. "A smart phone of the sort taken from Riley was unheard of ten years ago," he wrote. Technology is evolving too quickly for a single case to set precedents that will still be relevant thirty years from now. Imagine attempting to apply a court ruling on cassette tapes or mini disks (remember those?) from twenty years ago to a case today involving digitally downloaded songs. Digital downloads will be no more applicable in thirty years than these technologies are today.

As with most Supreme Court decisions, the rulings in these cases are still fairly narrow. But they are a portent of future debates. Chief Justice Roberts notes the near impossibility of an analogy that would allow the rules for a physical object familiar to our past to be applied to a digital one, consistent with our present. This is the future we have been previewing—one in which digital and analog blur to create an entirely new reality. Yes, these cases ruled on digital privacy. But they also offer implicit suggestions on how courts should approach other aspects of our daily lives that are impacted by the use of technology and the digitization of data. These cases—and the hundreds that are sure to follow them—foreshadow how our digital lifestyles will put to the test long-held values and social norms. When we consider that our homes and the potentially thousands of connected objects in it will soon contain extremely personal information—information that will be far more sensitive than what is held in today's smartphone—then we can begin to appreciate the Court's long view on this issue.

And yet the court's opinion in *Riley* occasioned mostly just obligatory mention by the media, which was far more interested in what the Court had to say about Hobby Lobby and its defiance of federal healthcare regulations. Not to dismiss the more popular case as without merit; however, given where our culture is headed, *Riley* is sure to have far more impact on everyday Americans. *Riley* is a foundational ruling. There has been the sense that "of course the police can't search my phone," but until *Riley* that wasn't legally true. Indeed, *Riley*'s legacy will be seen one day as the beginning of the great debate over how we interact with digital data and with one another over the next fifty years of digitization and connectivity.

That great debate won't be confined to questions about law enforcement or privacy either. It will include policy questions over the very framework of digitized data, such as how we access it, who else may access it, how it is allowed to grow and mature, and how providers manage it on our behalves. These are the issues we'll look at in this chapter.

A FRAMEWORK FOR PUBLIC POLICY

The strict parameters we once used to define privacy are evolving in a world increasingly defined by our dual existence in both the physical and the digital realm. Old laws that stated simply "Don't steal" made sense in a world in which your physical possessions were the only thing that anyone could steal. But what about our digital possessions? Do our online browsing, shopping, and Facebook page qualify as "possessions," off limits to anyone but ourselves? How far can the doctrine of property rights be extended into the digital realm, and does it cover the digital bread crumbs we leave behind? A new paradigm is emerging to govern our new digital-physical world. It could take a number of shapes, but one thing is clear: this new paradigm will be much more fluid—and therefore it will require a public policy approach that is much more fluid.

That is precisely why the 2014 *Riley* Supreme Court ruling is so consequential. The nine men and women who are meant to decide the Constitutional questions of our time have essentially ruled—unanimously in this case, a rarity for them—that it's way too early to make sweeping decisions about the intersection of personal property and digital data. While some privacy advocates might express a desire to have the Court create a strict legal framework, some fundamental realities about our new age make that impracticable.

For one thing, to live in this new world is to be a part of digital data. Put another way, when everything is digital, part of *you* is also digital. And when part of you is digital, that means others can "see" a part of you. A Pew survey found that 59 percent of Internet users do not believe that it's possible to be entirely anonymous online.[1]

If you have a cell phone, you aren't anonymous. If you have an email address, you aren't anonymous. If you have ever used your credit card to purchase a product or service, you aren't anonymous.

I stress this point because as a society we need to come to terms with what *is* possible in our new digital age. We have to accept that the great new things digital data brings us—and there are a lot of great things—come at the expense of complete and full anonymity. To enjoy even the barest sliver of these things, one must be a part of digital data; and to be a part, however small, of digital data is to give up your anonymity.

Which is not to say that there isn't a balance to be found. The Supreme Court will continue struggling to find that balance. The key point is: transformative conditions create fresh realities. Digitization—what is digitized and how the newly digitized data is subsequently used—is a fluid process, and the appropriate way to address fluidity from a public policy point of view is to implement adaptive mechanisms such as self-regulation.

People are just beginning to be fully aware of our concurrent existence in a dual digital-physical environment and to understand the implications thereof—spurred by numerous data breaches and the Snowden revelations. A new digitized world is exploding upon us rather suddenly. Even more, it's expanding at a quick rate, pushing its way into ever more intimate corners of our once-private lives.

One of the biggest policy issues confronting digital data, the one that all Americans—indeed all digitized data users everywhere—are more concerned about than ever before is privacy. Privacy in a digital world is essentially control over one's own digitized information. To avoid conflating the distinct concepts of government surveillance and private-sector data collection and aggregation on behalf of end-users, we will attempt to consider each independently while looking at both through the proposed framework of adaptive mechanisms.

Businesses and Data: As digitization seeps into and eventually consumes an increasing number of industries, more data will be digitized and as a result more businesses will have digital data that is relevant to each of us. Businesses retain information on behalf of consumers for really one primary reason—to improve the products and services they provide individuals. The real-world examples are multiplying. Netflix can offer better movie recommendations by knowing about your viewing habits. Fitness trackers from companies such as Fitbit and Jawbone can use personal data to create more appropriate fitness recommendations. Apple, Google, and others can create tomorrow's digital assistant by knowing as much about you as possible. Likewise, services like Foursquare and Yelp can offer more helpful recommendations with a richer understanding of your preferences. From retaining credit card information to facilitate seamless transactions to storing a host of personal digital data to provide a more customized experience, companies today are warehousing more data than ever before. And as more can be done with digital data, more data will be digitized and stored by companies.

Businesses are well aware of the risks and rewards associated with utilizing personal data and the risks uniquely associated with digitized data. We've seen how the scale of digitized data differs from that of analog data. Here's a simple example: I recently had several credit cards stolen from my wallet. I called the issuing banks immediately to cancel them, but the damage was done—thousands of dollars of merchandise had already been purchased illicitly at Best Buy and Apple stores not far away (at least the thieves had good taste). This is a not uncommon experience for anyone who has had a wallet stolen. But here's the point to my story. The theft was isolated to just me. Only my wallet was locked away in my truck. The digital parallel would mean 10 million wallets locked right next to mine. A digital data breach is on a much larger scale.

Companies recognize that the scale involved means missteps impact incredibly large portions of their most important customers. This is one of the reasons that self-regulation is so self-reinforcing. Companies should be afforded the maximum flexibility to choose how they apply privacy and security features in their products. Imposing strict government-mandated privacy regulations will only impede a company's ability to innovate and provide a unique user experience. Companies understand that they risk losing consumers' trust when apps fail either to protect data or to communicate effectively how the companies will use that data. The media is quick to jump on improper or veiled uses of data, which further drives the wedge between consumers and companies. These facts all help reinforce the motivation companies have to produce the outcome consumers and policy makers both want—innovative solutions and services together with strong privacy controls.

The spirit of transparency should govern how businesses use their stores of "big data." It is incumbent on private entities to pursue a responsible code of ethics when dealing with large amounts of personal data—which will only grow larger year after year. One large-scale misuse of big data could be enough to send the politicians and regulators flying to the cameras promising strangling regulations that will stifle further innovation. The onus, then, is on businesses to pursue industry-backed guidelines that promote best practices when it comes to big data.

Efforts should be made to increase consumer knowledge about existing privacy protections as well as about the benefits to consumers made possible through electronic data collection. Innovations in technology made possible through the collection and utilization of digital data should also be protected and promoted.

At the same time, a national standard on data breach notification is sorely needed. Right now, states have jurisdiction to create data breach notification laws. Although these laws have common elements, their differences create administrative problems for entities that operate nationally. This is one case where a single, comprehensive federal standard would benefit both consumers and businesses, which could create a single notification plan in the event of a data breach.

Before overarching privacy legislation or regulation is adopted, the definition of harm resulting from the use of consumer data collected electronically should be agreed to, including its standard of proof. We've seen many of the numerous potential benefits, both now and in the future, to consumers through the digitization and ultimately the collection of wide swaths of data. Public policy should protect these benefits—including by protecting the businesses whose innovation efforts would only be stymied by onerous government mandates.

International agreements, such as the US-EU Safe Harbor frame-work, allow companies operating in both the EU and the U.S. to transfer personal data between both international markets. Without agreements like this, multinational corporations would need to uti-lize other more onerous methods to comply with EU privacy laws. Agreements like these minimize marketplace friction and allow companies to more seamlessly operate across geopolitical borders.

We are only just beginning to unlock the potential for providing a more customized, and arguably more valuable, experience for the consumer. But the risks associated with collecting and storing per-sonal data are real. Companies should provide consumers with strong, reasonable, and contextually appropriate options to set their privacy and control their security. Individuals should have ultimate control and say over their data, and companies should be transpar-ent with regard to individual data. In the end, self-regulatory approaches to privacy protection should be encouraged and embraced. Innovation, privacy, and security are not mutually exclu-sive goals. In fact, in a well-functioning digital marketplace they are necessarily harmonious.

Government and Data: The revelations of the National Security Agency's data collection efforts, disclosed by Edward Snowden, have sparked a national debate about the degree to which the U.S. govern-ment—or any government—should be collecting data on its citizens. Naturally, the Snowden affair has generated a lot of congressional action in Washington, some good and some bad. In the world that we live in, where terrorists have the same access to digital data tools as anyone else, we cannot hope that analog intelligence efforts will keep us safe. That's why we should encourage legislative and regulatory reforms that place clear and reasonable restraints on governments' ability to collect data about citizens, while conceding that those efforts are critical for national security.

At the same time, there is a growing debate about the circumstances under which law enforcement agencies should be able to get information about citizens' locations. Unmanned aircraft systems, devices that spoof cell towers, and GPS tracking technologies have made it much easier for law enforcement to know the exact location of an individual.

The Electronic Communications Privacy Act (ECPA), which was passed in 1986, dictates how the federal government can obtain citizens' electronic communications. However, because the law was passed before cloud computing and other Internet technologies existed, its privacy protections are anachronistic. For instance, it allows law enforcement to obtain communications such as email with a mere subpoena, instead of a warrant, when email is stored in the cloud for more than 180 days. Many agree that aspects of ECPA no longer make sense and should be reformed to provide stronger privacy protection, which will increase consumer confidence in service providers.

This is a new frontier that we are entering, with no shortage of new questions bound to arise. Public policy will seek to provide answers to those questions, and the public policy framework should be clear. If regulation is the answer (and our first instinct should be that it is not), its primary goal should be to enhance individuals' continued use of technology. Policy responses to issues arising in the data-defined digital age should seek to enhance our trust in technology—not drive us from its use. Ideally, any policy responses should also be technologically neutral. No particular solution should be mandated over another, nor should any one element of the tech ecosystem be burdened with providing the sole solution or have the sole obligation to comply.

INCREASED SPECTRUM AND MOBILE BROADBAND USE

Another policy issue that will need attention is the topic of unlicensed spectrum. To give a bit of background, the Federal Communications Commission regulates the use of the radio frequency spectrum, through which over-the-air broadcasts, WiFi, Bluetooth, and mobile signals are sent. Users must obtain permission before transmitting to ensure that interference is properly managed and to prevent inefficient spectrum usage. The FCC manages the commercial spectrum using two broad categories: licensed and unlicensed. Unlicensed spectrum, which is open to anyone to use, is utilized in diverse ways—from garage door openers to baby monitors to smart meters. Indeed, CEA estimates that, solely in terms of the sale of devices using unlicensed spectrum to end-users, unlicensed spectrum generates over $62 billion per year in Incremental Retail Sales Value (IRSV). Just a few of the common consumer devices relying on spectrum for digital data distribution and reception now see annual shipment volumes surpassing 300 million devices a year. Unlicensed spectrum also powers much of the great innovation effort we've seen in recent years. Tech startups rely on unlicensed spectrum to test and hone new devices.

To ensure that entrepreneurs, startups, and consumers have the unlicensed spectrum they need to continue innovating and benefiting from digitized data, we not only need to protect the amount of unlicensed spectrum currently available, we also need to increase it and find new and more efficient uses for it. More liberal spectrum policies, better informed principles of spectrum rights, and an acceleration to repurpose spectrum from legacy systems will all help ensure a future in which digital data can move more freely.

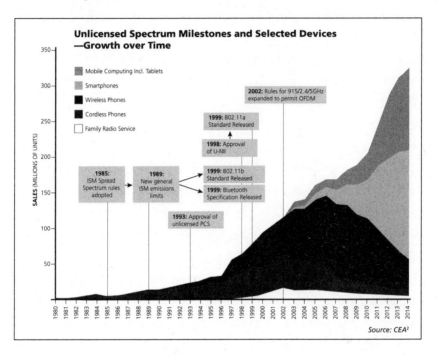

Unlicensed Spectrum Milestones and Selected Devices —Growth over Time

Source: CEA[2]

OPEN INTERNET

A heated topic that has galvanized a significant portion of the tech community is the idea known as open Internet, sometimes called "net neutrality." Briefly, proponents of net neutrality fear that Internet providers, to gain an economic advantage, will block or discriminate against Internet traffic that competes with their own service offerings. By these actions, proponents of net neutrality argue, Internet providers would be destroying the essentially "free" nature of the Internet, where all information is equal.

However, proponents miss a fundamental reason for the Internet's revolutionary success—namely, the competitive pressures that have worked to keep the Internet a relatively level playing field for all users. These pressures guaranteed that there are multiple Internet providers from which consumers can choose, but that's not all. Media and watchdog scrutiny, matched with consumer expectations, also provides a level of competitive pressure that precludes an Internet provider from discriminating against or blocking Internet traffic. Indeed, this natural competitive pressure works much in the way federal regulations would work, but without the innovation-stifling mandates.

Regulators should not pick winners and losers on the Internet. The best solution would be to let the market decide what consumers are willing to bear when it comes to paying more for important and time-sensitive services. Besides, letting government in the door is a dangerous game. Allowing regulations of one area of the Internet will only spur lawmakers to increase government's presence in other areas of the Internet—with overt government regulation that even net neutrality proponents would balk at. Healthy and open competition spurs innovation. It should also help craft an environment with the type of characteristics we want.

Regulators should focus on forward-looking efforts such as deploying more spectrum and promoting competition, an approach that will naturally maintain Internet openness. An open Internet spawns investment in broadband, which in turn allows data to move more freely where it can find its path to creating the most value— fulfilling the nature and highest potential of its existence.

FIRST, DO NO HARM

When considering policy prescriptions for some of the largest concerns affecting our digital destiny, lawmakers and regulators (as well as businesses) should abide by the Hippocratic principle of "First, do no harm." The Internet and its offspring—mobile phones, tablets, and the coming Internet of Things—have developed as the most dynamic wealth-creation engine the world has ever seen precisely because they have been allowed to grow organically.

This is not to say that the Internet and its growing influence on our lives do not bring concerns, particularly about security and privacy. All great leaps forward propelled by human ingenuity have had their dark sides—our digital destiny will be no different. But rather than panic in the face of these challenging obstacles and questions, we should confront them with a sober assurance that what has built the Internet will also sustain the Internet. There are some things that we have left behind in our old analog world—complete anonymity is one of them. Whether one regards the loss of anonymity as a bad thing, one cannot disregard the great gifts the Internet and digital data have provided and will provide. In the exchange, humanity has come out ahead.

CHAPTER 15

The Year Is 1450...

"We are drowning in information, while starving for wisdom."
—Edward O. Wilson in *Consilience: The Unity of Knowledge*

I n Greek mythology, the Titan Prometheus steals fire from the gods to give to humanity. While there are multiple versions of the story, they all note how Prometheus's theft supplied mankind with the essential element to begin civilization. The technology that Prometheus gave set human beings above the level of the animals, but only in technical skill. Wisdom was another matter. Plato's *Protagoras* expands upon the basic story, arguing that Prometheus's gifts gave man "the wisdom necessary to support life, but political wisdom he had not; for that was in the keeping of Zeus, and the power of Prometheus did not extend to entering into the citadel of heaven."

We should not take Plato's phrase "political wisdom" literally in this case. It has nothing to do with passing legislation or getting elected. Rather, Plato is referring to virtue, which in classical Athens was a political ideal. The problem is that Prometheus's great gift to mankind did not come with a manual. Fire could be used to warm the home, cook food, and smelt metal into tools; but it could also be used to burn, destroy, and produce weapons. It did not make man *better*; we were no closer in our attempt to reach the "citadel of heaven."

Likewise, our digital destiny doesn't inculcate virtue. Neither did Gutenberg's printing press or the invention of writing. On the question of right and wrong, data is silent. Much like earlier data revolutions, digital won't solve all of our problems. Certainly it will lead to great change. It will tear down barriers to progress and efficiency. It will alleviate suffering and disease. It will give a voice to the oppressed and knowledge to the ignorant. It will catapult mankind far beyond anything we could have imagined just a few decades earlier.

It just won't tell us how to live. On that question we are no further than the Greeks, whose insights from the myth of Prometheus hold true thousands of years later. In the form of digital data, we have been given a great gift. Nothing will ever be the same. Yet this gift came without instructions. How we choose to use it will decide whether our digital destiny is a joyous one or one filled with misery.

WHAT IS NEEDED

Technology influences our decision-making, which in turn affects human agency. We should see digitized data not merely as an output but as an active agent in our lives ahead. As Peter-Paul Verbeek notes in *Moralizing Technology: Understanding and Designing the Morality of Things*, "Life has become unthinkable without sophisticated technology. Contrary to what many people intuitively think, these technologies are not simply neutral instruments that facilitate our existence. While fulfilling their function, technologies do much more: they give shape to what we do and how we experience the world. And in doing so they contribute actively to the ways we live our lives."[1]

Unlike other morally ambiguous technologies such as atomic energy, digital data works alongside the human as a complementary force. Its function is dependent on human needs, inefficiencies, and barriers. It reveals our world in all its multiple dimensions and by doing so enlightens us on where it might be of use.

Technology and digitized data will not determine society's path. But we must recognize that it is more than just an instrument of which we have complete control. Data—and especially the combinatorial power of diverse streams of data—will reveal things we couldn't readily observe, and these things in turn will influence us to take action. In this way, digital data controls us as much as we control it.

But the things digital data reveals erase barriers between thought and action and lend themselves to achieving better ends. While we can envision a militarized drone becoming a more effective killing machine than a human-piloted jet, a driverless car will utilize data to save thousands, if not millions, of lives every year. Whereas governments can use digital data to exert greater control over citizens, citizens in turn can communicate, share, and organize to thwart state control. While we seem ever more dependent on our digital devices, digital data is also unchaining us from the tasks and chores that keep us from spending more time with our friends and family. Indeed, it is strengthening those relationships as well, removing the barrier of distance from our interactions.

Rarely in our fiction and entertainment do we look at the future through a lens of progress. We usually envision something far bleaker, dystopian, a time when either mankind has exploited technology toward destructive ends, sending us all back to the proverbial Dark Ages, or technology has become our masters. Perhaps the reality just isn't entertaining enough. Our likely future will be one in which the technologies powered by digital data will fit into our lives like a puzzle piece. No, not all will be happy days; but, by and large, we will use digital data—and it will use us—where need provides the greatest fit.

Need is a powerful force. It can lead to destructive ends, as our need to end World War II led to the development of the atomic bomb. But more often, fulfilling our needs improves our lives. Need drove Gutenberg to find a better way to replicate the written word. And need is driving our current innovations—finding the order in the chaos. Need pushes us to find better ways to deliver food to the hungry. Need helps us see the barriers in our healthcare. Need shines a light in our pursuit to find the better way.

Will we be better off? Will we be worse off? That is still an open question, but not a completely ambiguous one. Certainly we have benefited from the digital transformation that we've already experienced, but we have also seen great challenges emerging from it— from spam to data breaches to cyberbullying on social networks.

As we digitize greater swaths of our physical environment we will see completely new and previously unconsidered challenges. Need will help us find a path through the chaos. At the same time, we will receive the benefit of new services that we never imagined possible. The future looks bright. We will eventually look at these unimaginable innovations and wonder how we ever lived without them.

In our march forward we will influence the digital environment and digital data will continue to influence us, how we live and communicate, and ultimately who we are. In the end we will get to decide what our destiny looks likes in this new paradigm.

THE NEXT PHASE

It is worth asking what the next phase of our digital transformation will look like. Where will need drive us to use digital data next? Isn't that what this whole book has been about, you ask? Yes, but I'm speaking about our near-future, the months and years immediately ahead. Curating that data, sifting it, finding the gems that allow us to remove yet another barrier will be where the next great innovations will occur.

As Edward O. Wilson wrote in his 1998 book *Consilience*, "Thanks to science and technology, access to factual knowledge of all kinds is rising exponentially while dropping in unit cost. It is destined to become global and democratic. Soon it will be available everywhere on television and computer screens. What then? The answer is clear: synthesis. We are drowning in information, while starving for wisdom. The world henceforth will be run by synthesizers, people able to put together the right information at the right time, think critically about it, and make important choices wisely."

Indeed, digital data will change the way we live, and not just the basic functions of living. It will change every aspect of our daily lives: how we live, work, and communicate. Mobile phones provide a modern and contextual example. They've given rise to new forms of communication and even helped the development of new "languages" as texting and other messaging services have evolved. Smartphones redefine geographic space; now you can have a "live" conversation with someone who is miles away or shop at a retailer while you are physically standing within the confines of a competing store. Space has shrunk to the size of our mobile phones. Digital data, in its many manifestations on the horizon, will further shrink distance in all its varied forms as we isolate the valuable data—the gems—from the noise.

As the Greeks connected the stars in the night sky to find patterns, shapes, and things, so too we will look at the stars in the digital universe and create something out of nothing. This is one part of our digital destiny with which we have little direct experience as of yet. Our devices—from smartphones to laptops and PCs—are already becoming "smarter," in that they better understand what we need; but what we see today will pale in comparison to what we will make possible tomorrow.

The fact is that most of us still understand the Internet as a single entity—a giant labyrinth one enters from thousands of different points. From there, we scrape through the darkness grasping at these valuable points of light. Search is just the most advanced way we have devised to make sense of the Internet. But, as I've argued, search is failing us even now, as the digital universe expands through the billions of billions of connected devices that are coming our way.

Using search as we understand it to navigate this abyss won't just be more difficult; it will literally be impossible. The whole nature of the Internet will shift from a community of web sites to an ocean of data. But through our devices and other digital objects, we will be able to find the lights. We just won't have to do it ourselves. Our devices and other digital objects will deliver a curated experience providing settings that fit our moods, recommendations that meet our needs, and the distinctiveness of an experience perfectly mea-sured for the individual. This curation is made possible only through the five pillars of digitized data that I laid out in the earlier chapters of this book—ubiquitous computing, explosion of digitized devices, universal connectivity, digital data storage, and sensors.

If we step back and look at our lives abstractly, we see that decisions we make are not made exclusively by us in isolation, but are shaped by the interactions we have with the technology we use. Most modern cars either won't start until the seatbelt is engaged or will provide an (often painfully annoying) indication that the seat belt isn't engaged after the vehicle is started. Either we engage the seat belt and stop the incessant beeping or we "endure" the beeping and continue to drive without the seat belt engaged. The technology exerts its influence on us in either case.

The influence of technology is growing, as we already see from the devices of today. Our devices empower us to make decisions and choices that were once unavailable. For instance, technology allows us to contact someone whom we might otherwise not have contacted. That's technology influencing our decisions. Our future, and the decisions we make then, will see even more influence from the digitization of data that is now taking place—but unlike in my seatbelt example, much of this influence will be less visible to us. The digitization of data began with revealing unobservable data, making the invisible visible. That is but the first step. Our digital destiny will fully take hold when this newly visible data fades into the background and begins making decisions on our behalf. Once again data will seem invisible, but the influence of digital data will be immense.

THE DEVIL IN THE DETAILS

While our digital destiny will bring significant progress to our daily lives, it will never be perfect. Researchers have shown that we can predict patients going into cardiac arrest with up to a four-hour lead time, but they have also found false positive rates of 20 percent. The question in a data paradigm is how high of a false positive rate will we tolerate before we are willing to take action. Is it 20 percent? 10 percent? 5 percent? These are the important decisions before us as we move towards our digital destiny.

Data is marching on—in a long cadence that has been going on for centuries. We have often thought of ourselves as the puppeteers. I argue here that as much as we are pushing data, digital data is also pushing us. The things discussed in this book aren't magic. They are part of a continuum—a long continuum of evolution that began when man first tried to express a thought or describe what he saw. The never-ending cycle of chaos and order was present at the beginning of history and will be there until the bitter end.

The cycle will continue because the properties of data are absolute. Data has always been pushing for release, for liberation. It pushes us to find better ways—sometimes faster, sometimes clearer, sometimes louder—of unchaining itself from its physical boundaries of space and time. In the process, data tears down barriers, generating a massive amount of chaos as it is unleashed. But in our struggle to make sense of the data explosion, we find the order—we filter and arrange and sort. We are also led to the order because we follow the data.

Digital is the ultimate evolution of data's long history of liberation. Yet we must not think that digital is the end of that history, nor that we will ever reach the end. Mankind can never know all there is to know in the universe at the precise moment he needs to know it—but we can get closer. In our striving to close that gap, we will encounter many obstacles. These will limit our understanding of what the data is trying to tell us. We will be led down dead ends. Some journeys will turn up nothing.

Fortunately, we're quite adept at wallowing in the details. The details of digital data sometimes appear like great mountain chains. How will we ever feel secure when our identities are in the crosshairs of hackers? How can we ever find comfort when our children are subject to such derision and cruelty online? In a world of so much poverty and suffering, what does a sensorized urinal really mean in the grand scheme of things? Where's the benefit to the man who just lost his job to a machine?

These details, the demons of our new world, hardly appear like mere details, particularly to those who are affected by them the most. But details they remain. And how we find our way through the devilish details will be part of our digital destiny as well. That journey will not be clean or painless. It will be hard; it will create great controversy. It might even lead to conflict.

Yet we must make this journey because we cannot undo digital data. Just as our ancestors could not forget about how to make fire—they couldn't just give it back to Prometheus—we have to learn to live in this new world. Perhaps we are a bit wiser than our forebears who experienced the great tumult unleashed by the printing press. Our accumulated wisdom will help us identify the lights.

THE FULFILLMENT OF SELF

Our digital destiny is a story about personalization and customization. It is about allowing data to fulfill the nature and potential of its existence by enabling us to realize the potential of ours. By far the greatest consequence of digital data will be the ability of the individual to fulfill his needs and desires to an unprecedented degree.

In economic terms, this will be the equivalent of Henry Ford's ability to make his cars affordable to the common man. We all hear about the miraculous things scientists and researchers are doing these days, but until we see—directly, not in an abstract sense—how they will impact our lives, we are no better off than before.

Much of the great personalization and customization promised by digital data is still before us. Indeed, we are in the great swirling chaos of the data cycle. We will likely not find the order for some time. But that doesn't mean we won't begin to see little corners of our lives affected by digital data come into better focus. The ability to shop from home; the ability to telecommute; the ability to get personalized news, weather, sports, and other information—all of these things that we already take for granted are just the tip of the iceberg for the fulfillment of self that will come with digital data.

In the not-so-distant future, we will be able to customize the world to our lives, rather than the reverse. This stunning development is nothing less than a complete revolution in human affairs. For millennia, human beings have had to conform to their surrounding—their circumstance, their "lot in life." We will always do so, but our "lot" will become smaller and smaller. Put another way, the things out of our control will become fewer and fewer. We will exercise a level of control over the direction of our lives that the kings and emperors of history would not have been able to fathom.

Even at the farthest end of the economic scale, the poor will have access to knowledge, to cheap goods, to freedom from certain adverse circumstances. Those of us in the middle will find that we can shape our lives according to our wishes.

Is this too much to hope for? Perhaps. But we always must remember that with our new-found freedom will come unforeseen challenges. We cannot expect to remain as we are in our digital future. As Peter-Paul Verbeek has noted, "with the development of ambient intelligence and persuasive technology, technologies start to interfere openly with our behavior, interacting with people in sophisticated ways and subtly persuading them to change their behavior."[2] The digitization of data will allow these sophisticated interactions and behavior changes to take place at an increasing rate in our lives. A key storyline of the great digitization we are undergoing is the influence technology has on us and the decisions we make throughout the day.

Some of these decisions will be in accord with our needs and desires; others will not. It will not be roses to the end of our days, but nor will it be the dreary dystopia of so many science fiction stories. It will be on the whole better, but we will always remain imperfect, and human.

THE YEAR IS 1450 ...

In *The Signal and the Noise: Why So Many Predictions Fail—but Some Don't*, Nate Silver writes that digital data will "produce progress—eventually. How quickly it does, and whether we regress in the meantime, will depend on us." As a pioneer of the next generation's data scientists, Silver has helped unlock the great potential in digital data. Indeed, digital data has made Silver a very successful man. Yet even Silver understands data's limitations.

Throughout this book I have tried to temper my genuine excitement and hopefulness about our digital destiny with the grim realities. I have not wanted to dwell on the many violent upheavals that were spawned by previous data revolutions, but we cannot discuss our future without acknowledging both sides. On the whole, we should look at our digital destiny as the great era in humanity it will be—an era of solutions, of abundance, of creativity, and of less suffering. But problems, scarcity, and suffering will also still exist along the rocky road to this next era.

Again, we must remember that there is a historical precedent for the period in which we find ourselves. It's 1450 and Gutenberg's printing press is before us for the first time. The time for progress is here. There is more going on with data than ever before. We are entering a new paradigm that will require an entirely new way of thinking about the world and our place in it.

It's 1450 and the explosion of knowledge and self-fulfillment is about to rock age-old institutions. Some of those institutions were tottering anyway, and they will fall with barely a whisper; others stand tall, seemingly impervious to any change. But nothing will remain unchanged in our digital destiny. Those institutions will fall just as so many others have throughout history, and their destruction will bring great violence and suffering, but also wisdom and truth.

It's 1450 and the world stands on the precipice of a dark chasm. There is light on the other side, brilliant, blindingly clear light. It is our digital destiny beckoning us to take the plunge, to have some faith in our ability to find the right path, as we've done in ages past. It's calling us to take one more step. And so we do, knowing that nothing will ever be the same.

Acknowledgments

In October 2012, Gary Shapiro and I shared a lunch over which I told him about the manuscript I had been working on since January 2011. Sight unseen, he embraced the project and committed his support. He was the first to read it in its entirely—something he did while on a rare vacation with his family. I'm deeply grateful to him for his enthusiasm and encouragement, his commitment and support. I'm privileged to have had him write the foreword to this book.

I was lucky to work with two extremely gifted editors. Blake Dvorak helped me work through every inch of the manuscript. For this he deserves more than a gracious thank you for the many late nights he tolerated me, expounding and pontificating, on the themes of this book. Not only a gifted editor, along the way, Blake taught me valuable lessons about writing that will stay with me for a lifetime. Elizabeth Kantor has an amazing ability to take something I reworked a myriad of times and in a single sweep, offer a subtle suggestion that turns words into prose. Elizabeth was painfully thorough and always pushed me for precision and accuracy. These two are literary masters.

I'm thankful for comments and suggestions from Brian Markwalter, Sean Murphy, Chris Ely, Dave Wilson, Sean Parker, Michael Petricone, Jeff Joseph, and Danielle Cassagnol. I want to especially note the invaluable feedback I received from Alex Reynolds and Michael Bergman. Michael's feedback was so thorough that in many instances I incorporated his comments directly into the text of the manuscript. He pushed my thinking and writing and for that I am grateful.

I am blessed to work with a brilliant team of researchers who run CEA's Research Center. It is a boutique, tech-focused library that is the premier source for industry information and market intelligence. They are a sophisticated group of experts whom I rely on heavily. I accost them daily with requests. Richard Kowalski, Angela Titone, and Katherine Rutkowski—thank you for all you do!

Susan Littleton will never know just how much I adore and appreciate her. Her ability to design and execute a plan is second to none. I can't think of a single thing I wouldn't turn over to her to implement and accomplish. I've learned it takes the proverbial village to turn a manuscript into a book, and Susan is the mayor, police superintendent, and fire chief for the village that came together in support of this project. Michael Brown is the smoothest of operators who transforms stress and deadlines into actionable results. CEA has the most talented creative team I have ever witnessed—made up of John J. Lindsey II, Ian Shields, Collin King and Matt Patchett. Multiple times they took an opaque idea and translated it into something tangible. I thank Jenni Moyer, Victoria Velez, Tina Anthony, Don Schaefer, Kasey Stanton, Laura Hubbard, and Pam Golden. Johannah O'Keefe, I could write an entire book about all you do—you organize and juggle and keep everything on track. You are the one I fear isn't credited enough.

Thanks to the team at the Pinkston Group—especially Christian Pinkston and David Fouse—who have provided valuable guidance throughout.

Thank you to Glenda MacMullin for her continuous support and advice throughout this project. And my deep gratitude to Tyler Suiters and Rachel Horn, two of the most gifted thinkers and writers I know.

Last, I appreciate the hundreds of people along the way who have challenged my thinking and forced me to solidify my views. I look forward to our next discussion.

A funny thing happens when you write a book. You spend the first few years completely alone and isolated, which are only to be succeeded by months where you are surrounded by dozens of people working to fix everything you didn't do (right) in the first few years. I am forever indebted. Thank you!

Notes

INTRODUCTION

1. "Early Estimate of Motor Vehicle Traffic Fatalities in 2013," Traffic Safety Facts, U.S. Department of Transportation National Highway Traffic Safety Association, May 2014, http://www-nrd.nhtsa.dot.gov/Pubs/812024.pdf.

2. Larry Copeland, "Traffic Fatalities Increased 3.3% in 2012," *USA Today*, November 14, 2013, http://www.usatoday.com/story/news/nation/2013/11/14/highway-fatalities-2012/3528665/.

3. "Road Traffic Deaths: Data by Country," World Health Organization Global Health Observatory Data Repository, 2014, http://apps.who.int/gho/data/node.main.A997.

4. Joanna Brenner, "3% of Americans Use Dial-up at Home," Pew Research Center FactTank: News in the Numbers, August 21, 2013, http://www.pewresearch.org/fact-tank/2013/08/21/3-of-americans-use-dial-up-at-home/.

5.   Michael DeGusta, "Are Smartphones Spreading Faster Than Any Technology in Human History? Mobile Computers Are on Track to Saturate Markets in the U.S. and the Developing World in Record Time," *MIT Technology Review*, May 9, 2012, http://www.technologyreview.com/news/427787/ are-smartphones-spreading-faster-than-any-technology-in-human-history/?ref=rss.

6.   Kevin Ashton, "That 'Internet of Things' Thing: In the Real World, Things Matter More Than Ideas," *RFID Journal*, June 22, 2009, http://www. rfidjournal.com/articles/view?4986.

CHAPTER 1: THE BEGINNING OF OUR VOYAGE AND THE PROPERTIES OF DATA

Epigraph. Russell Fox, Max Gorbuny, and Robert Hooke, *The Science of Science: Methods of Interpreting Physical Phenomena* (New York: Walker, 1963), 51.

1.   Arthur C. Clarke, *Profiles of the Future: An Inquiry into the Limits of the Possible*, rev. ed. (New York: Harper and Row, 1973), 21.

2.   Gary Shapiro, *Ninja Innovation: The Ten Killer Strategies of the World's Most Successful Businesses* (New York: William Morrow, 2013), 19.

3.   "U.S. Makes Ninety Percent of World's Automobiles," *Popular Science Monthly* (November 1929): 84, http://books.google.com/books?id=FSgDAA AAMBAJ&pg=PA84&dq=U.S.+makes+ninety+percent+of+world%27s+aut omobiles#v=onepage&q=U.S.%20makes%20ninety%20percent%20of%20 world's%20automobiles&f=false.

4.   U.S. Consumer Electronics Sales and Forecast, Consumer Electronics Association.

5.   Kevin Kelly, *What Technology Wants* (New York: Penguin, 2011), 4.

6.   Ibid, 15.

7.   Ibid, 270.

CHAPTER 2: THE SEEDS OF OUR DIGITAL DESTINY

Epigraph. Carl Gustav Jung, *The Archetypes and the Collective Unconscious* (Princeton University Press, 1959), 32.

1.   "Internet Sapping Broadcast News Audience," Pew Research Center for People & the Press, June 11, 2000, http://www.people-press.org/2000/06/11/ internet-sapping-broadcast-news-audience/.

2. "Key Indicators in Media & News," Pew Research Journalism Project, March 26, 2014, http://www.journalism.org/2014/03/26/state-of-the-news-media-2014-key-indicators-in-media-and-news/.

3. David Shedden, "New Media Timeline (2000)," *Poynter*, May 2, 2013, http://www.poynter.org/uncategorized/28786/new-media-timeline-2000/.

4. "The Internet News Audience Goes Ordinary," Pew Research Center for People & the Press, January 14, 1999, http://www.people-press.org/1999/01/14/the-internet-news-audience-goes-ordinary/.

5. "The Web at 25 in the U.S.," Pew Research Internet Project, February 27, 2014, http://www.pewinternet.org/2014/02/27/summary-of-findings-3/.

6. Steve Outing, "Crisis Notes from the Online Media," *Poynter*, September 2, 2002, http://www.poynter.org/uncategorized/2363/crisis-notes-from-the-online-media/.

7. Gary Shapiro, "Bin Laden's Death and the Information Revolution," *Forbes*, May 5, 2011, http://www.forbes.com/2011/05/05/bin-laden-death-information-revolution.html.

8. "The Rational and Attentive News Consumer," *American Press Institute*, March 17, 2014, http://www.americanpressinstitute.org/publications/reports/survey-research/rational-attentive-news-consumer/.

9. Howard Gardner, *The Mind's New Science: A History of the Cognitive Revolution* (New York: Basic Books, 1987), 144.

10. Gordon E. Moore, "Cramming More Components onto Integrated Circuits," *Electronics* 38, no. 8 (April 19, 1965): 83.

11. Data published on the websites of AMD, Apple, IBM, Intel, MOS Technologies, Motorola, Oracle, RCA, and Zilog.

12. David Friedman, "Steven Sasson, Inventor of the Digital Camera," *The Donut Project*, April 12, 2011, http://www.thedonutproject.com/inspiration/inventor-of-the-digital-camera/.

13. Erik Sandberg-Diment, "Personal Computers: Software for the Macintosh: Plenty on the Way," *New York Times*, January 31, 1984, http://www.nytimes.com/1984/01/31/science/personal-computers-software-for-the-macintosh-plenty-on-the-way.html.

14. "Sir Timothy Berners-Lee: Connecting All Humanity," Academy of Achievement: A Museum of Living History, June 22, 2007, http://www.achievement.org/autodoc/page/ber1int-1.

15. Timothy Berners-Lee, "World Wide Web—Summary," http://info.cern.ch/hypertext/WWW/Summary.html.

16. Jamie Reno and Michael Leverson Meyer, "HDTV Is Finally Here! Well, Almost," *Newsweek*, August 17, 1998.

17. Ray Kurzweil, *The Age of Spiritual Machines* (New York: Penguin, 2000), 36–37.

18. Hong Qu, "Social Media and the Boston Bombings," Nieman Reports, no date, http://www.nieman.harvard.edu/reports/article/102871/Social-Media-and-the-Boston-Bombings.aspx.

19. Dan Gilgoff and Jane J. Lee, "Social Media Shapes Boston Bombings Response," *National Geographic News*, April 15, 2013, http://news.nationalgeographic.com/news/2013/13/130415-boston-marathon-bombings-terrorism-social-media-twitter-facebook/?rptregcta=reg_free_np&rptregcampaign=20131016_rw_membership_r1p_us_se_w#.

20. Laura Petrecca, "After Bombings, Social Media Informs (and Misinforms)," *USA Today*, April 23, 2013, http://www.usatoday.com/story/news/2013/04/23/social-media-boston-marathon-bombings/2106701/.

21. David Montgomery, Sari Horwitz, and Marc Fisher, "Police, Citizens and Technology Factor into Boston Bombing Probe," *Washington Post*, April 20, 2013, http://www.washingtonpost.com/world/national-security/inside-the-investigation-of-the-boston-marathon-bombing/2013/04/20/19d8c322-a8ff-11e2-b029-8fb7e977ef71_print.html.

22. Ibid.

CHAPTER 3: WHEN DATA IS DIGITIZED

Epigraph. Terry Pratchett, *Lords and Ladies* (New York: HarperPrism, 1992), 1.

1. Mirren Gidda, "Edward Snowden and the NSA Files—Timeline," *Guardian*, July 25, 2013, http://www.theguardian.com/world/2013/jun/23/edward-snowden-nsa-files-timeline.

2. Michael B. Kelley, "NSA: Snowden Stole 1.7 Million Classified Documents and Still Has Access to Most of Them," *Business Insider*, December 13, 2013, http://www.businessinsider.com/how-many-docs-did-snowden-take-2013-12.

3. Scott Vaughen, "Counting the Permutations of the Rubik's Cube," http://faculty.mc3.edu/cvaughen/rubikscube/cube_counting.ppt.

4. SINTEF, "Big Data, for Better or Worse: 90% of World's Data Generated over Last Two Years," *Science Daily*, May 22, 2013, http://www.sciencedaily.com/releases/2013/05/130522085217.htm.

5. Martin Hilbert and Priscilla López, "The World's Technological Capacity to Store, Communicate, and Compute Information," *Science* 322, 60 (2011): http://www.sciencemag.org/content/332/6025/60.full.pdf?keytype=ref&siteid=sci&ijkey=89mdkEW.yhHlM.

6. IDC, "The Digital Universe of Opportunities: Rich Data and the Increasing Value of the Internet of Things," April 2014, http://www.emc.com/leadership/digital-universe/2014iview/index.htm.

7. IDC, "The Digital Universe in 2020: EMC Digital Universe Study with Research and Analysis by IDC," Digital Universe report, 2014, http://www.emc.com/leadership/digital-universe/index.htm?pid=home-dig-uni-090414.

8. IDC, "The Internet of Things: Data from Embedded Systems Will Account for 10% of Digital Universe by 2010," Digital Universe Report, April 2014, http://www.emc.com/leadership/digital-universe/2014iview/internet-of-things.htm.

9. Josh James, "Data Never Sleeps 2.0," Domosphere, April 23, 2014, http://www.domo.com/blog/2014/04/data-never-sleeps-2-0/.

10. "Cisco Visual Networking Index: Global Mobile Data Traffic Forecast Update, 2013–2018," Cisco, February 5, 2014, http://www.cisco.com/c/en/us/solutions/collateral/service-provider/visual-networking-index-vni/white_paper_c11-520862.html.

11. "Cisco Visual Networking Index: Forecast and Methodology, 2013–2018," Cisco, June 10, 2014, http://www.cisco.com/c/en/us/solutions/collateral/service-provider/ip-ngn-ip-next-generation-network/white_paper_c11-481360.html.

12. "Broadband vs. Dialup Adoption over Time," Pew Research Internet Project, September 2013, http://www.pewinternet.org/data-trend/internet-use/connection-type/.

13. Matt Komorowski, "A History of Storage Cost," mkomo.com, September 8, 2009 (updated March 9, 2014; see note below), http://www.mkomo.com/cost-per-gigabyte.

14. Matt Komorowski, "A History of Storage Cost," mkomo.com, March 9, 2014, http://www.mkomo.com/cost-per-gigabyte-update.

15. Matt Komorowski, "A History of Storage Cost (Updated)," mkomo.com, March 9, 2014, http://www.mkomo.com/cost-per-gigabyte.

16. Ibid.

CHAPTER 4: THE SENSORIZATION OF OBJECTS

Epigraph. Plato, "Laches," *The Dialogues of Plato* (London: Macmillan, 1875), 85.

1. "Data: Ethiopia," The World Bank, 2014, http://data.worldbank.org/country/ethiopia#cp_gep.

2. *The World Fact Book*, Central Intelligence Agency, June 22, 2014, https://www.cia.gov/library/publications/the-world-factbook/geos/et.html.

3. Paul Saffo, "Sensors: The Next Wave of Infotech Innovation: From 1997 Year Forecast," Saffo.com, no date, http://www.saffo.com/essays/sensors-the-next-wave-of-infotech-innovation/.

4. Mark Gurman, "iWatch's Novelty Emerges As Apple Taps Sensor and Fitness Experts," 9 to 5 Mac, July 18, 2013, http://9to5mac.com/2013/07/18/apple-stacks-iwatch-team-with-sensor-fitness-experts/.

5. Mark Fischetti, "IBM Simulates 4.5 Percent of the Human Brain, and All of the Cat Brain," *Scientific American*, October 25, 2011, http://www.scientificamerican.com/article/graphic-science-ibm-simulates-4-percent-human-brain-all-of-cat-brain/.

6. Matt Soniak, "How Did the Duck Hunt Gun Work?" Mental Floss, January 14, 2011, http://mentalfloss.com/article/26875/how-did-duck-hunt-gun-work.

7. Marc Perton, "The Oculus Rift 'Crystal Cove' Prototype is 2014's Best of CES Winner," Endgadget, January 9, 2014, http://www.engadget.com/2014/01/09/the-oculus-rift-crystal-cove-prototype-is-2014s-best-of-c/.

8. Nicole Lee, "NASA JPL Takes a VR Tour of Mars with Oculus Rift and Virtuix Omni," Endgadget, August 5, 2013, http://www.engadget.com/2013/08/05/nasa-jpl-oculus-rift-virtuix-omni/.

9. Perton, "The Oculus Rift 'Crystal Cove.'"

10. Jim Edwards, "Ford Exec: 'We Know Everyone Who Breaks the Law,' Thanks to Our GPS in Your Car," *Business Insider Australia*, January 9, 2014, http://www.businessinsider.com.au/ford-exec-gps-2014-1.

11. IDC, "The Internet of Things: Data from Embedded Systems Will Account for 10% of Digital Universe by 2010," *Digital Universe Report*, April 2014, http://www.emc.com/leadership/digital-universe/2014iview/internet-of-things.htm.

12. "EMC Digital Universe Study with Research and Analysis by IDC," 2014, http://www.emc.com/leadership/digital-universe/index.htm; "Infobrief: The Digital Universe of Opportunities," April 2014, http://www.emc.com/collateral/analyst-reports/idc-digital-universe-2014.pdf.

13. IDC, "The Internet of Things: Data from Embedded Systems Will Account for 10% of Digital Universe by 2010."

14. "EMC Digital Universe Study with Research and Analysis by IDC," 2014, http://www.emc.com/leadership/digital-universe/index.htm; "Infobrief: The Digital Universe of Opportunities," April 2014, http://www.emc.com/collateral/analyst-reports/idc-digital-universe-2014.pdf.

15. Walden Rhines, "Viewpoint: Is Semiconductor Industry Consolidation Inevitable?," *EETimes*, April 5, 2010, http://www.eetimes.com/document.asp?doc_id=1173451&page_number=2.

16. "Market Statistics," World Semiconductor Trade Statistics, multiple years, http://www.semiconductors.org/subscribers_only/bluebooks/.

CHAPTER 5: SECOND-ORDER EFFECTS OF DIGITAL DATA

1. Lee Rainie, Sarah Kiesler, Ruogu Kang, and Mary Madden, "Anonymity, Privacy, and Security Online," Pew Research Internet Project, September 5, 2013, http://www.pewinternet.org/2013/09/05/anonymity-privacy-and-security-online/.

2. "CNBC's Rick Santelli's Tea Party," The Heritage Foundation, YouTube, https://www.youtube.com/watch?v=zp-Jw-5Kx8k.

3. Carol Huang, "Facebook and Twitter Key to Arab Spring Uprisings: Report," *National* (UAE), June 6, 2011, http://www.thenational.ae/news/uae-news/facebook-and-twitter-key-to-arab-spring-uprisings-report.

4. Jessica Litman, "The Exclusive Right to Read," *Cardozo Arts & Entertainment Law Journal* 13 (1994): 29.

5. Lawrence Lessig, "It Is about Time: Getting Our Values around Copyright Right," 2009 Educause Congress, November 5, 2009.

6. "Factsheet on the 'Right to Be Forgotten' Ruling (C-131/12)," European Commission, no date, http://ec.europa.eu/justice/data-protection/files/factsheets/factsheet_data_protection_en.pdf.

7. Bill Keller, "Erasing History," *New York Times*, April 28, 2013, http://www.nytimes.com/2013/04/29/opinion/keller-erasing-history.html?partner=rss&emc=rss&_r=2&pagewanted=all&.

8. Ibid.

CHAPTER 6: IN THE YEAR 2025…

Epigraph. Hiroshi Inose and J. R. Pierce, *Information Technology and Civilization* (New York: W. H. Freeman, 1984), 210.

1. Melissa J. Perenson, "Microsoft Debuts New 'Minority Report'–Like Surface Computer," PC World, May 29, 2007, http://www.pcworld.com/article/132352/article.html.

2. Austin Carr, "Iris Scanners Create the Most Secure City in the World. Welcome, Big Brother," Fast Company, August 18, 2010, http://www.fastcompany.com/1683302/iris-scanners-create-most-secure-city-world-welcome-big-brother.

3. Richard Gray, "*Minority Report*–Style Advertising Billboards to Target Consumers," *Telegraph*, August 1, 2010, http://www.telegraph.co.uk/technology/news/7920057/Minority-Report-style-advertising-billboards-to-target-consumers.html.

4. Charles Arthur, "Why *Minority Report* Was Spot On: It's Only Eight Years Since Stephen Spielberg's *Minority Report* Amazed Audiences with Its Futuristic Technology. But Now Science Is Fast Catching Up," *Guardian*, June 16, 2010, http://www.theguardian.com/technology/2010/jun/16/minority-report-technology-comes-true.

5. Lisa Kennedy, "Spielberg in the Twilight Zone," Wired 10:06 (June 2002), http://archive.wired.com/wired/archive/10.06/spielberg.html.

6. Erik Brynjolfsson and Andrew McAfee, *The Second Machine Age: Work, Progress, and Prosperity in a Time of Brilliant Technologies* (New York: W. W. Norton & Company, 2014), 9–10.

CHAPTER 7: DRIVERLESS CARS AND THE NEW DIGITAL AGE OF TRAVEL

Epigraph. Tyler Cowen, *Average Is Over: Powering America beyond the Age of the Great Stagnation* (New York: Plume, 2014), 8.

1. Jeff Wise, "What Really Happened Aboard Air France 447," Popular Mechanics, December 6, 2011, http://www.popularmechanics.com/technology/aviation/crashes/what-really-happened-aboard-air-france-447-6611877-2.

2. John Croft, "Wiener's Laws," Things With Wings, Aviation Week, July 28, 2013, http://aviationweek.com/blog/wiener-s-laws.

3. "Fly-by-Wire for Combat Aircraft," Flight International, August 23, 1973, http://www.flightglobal.com/pdfarchive/view/1973/1973%20-%202228.html.

4. Ian Moir, Allan G. Seabridge, and Malcolm Jukes, *Civil Avionics Systems* (London: iMechE/Professional Engineering Publishing Ltd, 2003), 259.

5. Jad Mouawad and Christopher Drew, "Airline Industry at Its Safest Since the Dawn of the Jet Age," *New York Times*, February 11, 2103, http://www.nytimes.com/2013/02/12/business/2012-was-the-safest-year-for-airlines-globally-since-1945.html?pagewanted=all&_r=0.

6. See "Crash Rates per Year," Bureau of Aircraft Accidents Archives, http://www.baaa-acro.com/general-statistics/crashs-rate-per-year/.

7. Javier Espinoza, "Is the Driverless Tractor the Future for Farmers: Uptake Is Still Slow, but These Robots Could Change the Future of Agriculture," *Wall Street Journal*, November 20, 2013, http://online.wsj.com/news/articles/SB10001424052702303936904579177531155474254.

8. Alisa Priddle and Chis Woodyard, "Google Discloses Costs of Its Driverless Car Tests," *USA Today*, June 14, 2012, http://content.usatoday.com/communities/driveon/post/2012/06/google-discloses-costs-of-its-driverless-car-tests/1#.U8nMFPldWSq.

9. Arin Greenwood, "The Median Home Price is $188,900. Here's What That Actually Buys You," *Huffington Post*, March 14, 2014, http://www.huffingtonpost.com/2014/03/13/median-home-price-2014_n_4957604.html.

10. "U.S. Department of Transportation Releases Policy on Automated Vehicle Development," National Highway Traffic Safety Administration, May 30, 2013, http://www.nhtsa.gov/About+NHTSA/Press+Releases/U.S.+Department+of+Transportation+Releases+Policy+on+Automated+Vehicle+Development.

11. "Effects of Next-Generation Vehicles on Travel Demand & Highway Capacity," Fehr & Peers, no date, http://www.fehrandpeers.com/fpthink/nextgenerationvehicles/.

12. "When Will You Be Able to Buy a Driverless Car?" *Venture Beat*, 2014, http://venturebeat.files.wordpress.com/2014/07/autonomous_driverless_car_infographic_predictions.png.

13. "Preparing a Nation for Autonomous Vehicles," Eno Center for Transportation, October 2013, https://www.enotrans.org/wp-content/uploads/wpsc/downloadables/AV-paper.pdf.

14.   Ibid, 11.

15.   Associated Press, "L.A. Stoplights Synchronized to Improve Traffic," CBS News, May 25, 2013, http://www.cbsnews.com/news/la-stoplights-synchronized-to-improve-traffic.

16.   Clifford Winston, "Paving the Way for Driverless Cars," *Wall Street Journal*, July 18, 2012, http://www.brookings.edu/research/opinions/2012/07/18-driverless-cars-winston.

17.   Ibid.

18.   Bryant Walker Smith, "Automated Vehicles Are Probably Legal in the United States," Stanford Cyberlaw, November 1, 2012, http://cyberlaw.stanford.edu/downloads/20121101-AutomatedVehiclesareProbablyLegalInTheUS.htm.

19.   "Preparing a Nation," 11–14.

20.   Matt Burns, "Ex-Ford CEO Alan Mulally Joins Google's Board of Directors," Tech Crunch, July 15, 2014, http://techcrunch.com/2014/07/15/ex-ford-ceo-alan-mulally-joins-googles-board-of-directors/.

21.   "Preparing a Nation," 12.

22.   Ibid.

23.   Ibid., 13.

24.   National Highway Traffic Safety Administration, "National Motor Vehicle Crash Causation Survey," U.S. Department of Transportation, Report DOT HS 811 059, 2008.

25.   Cambridge Systematics, "Crashes vs. Congestion: What's the Cost to Society?" American Automobile Association, 2011.

CHAPTER 8: THE INTERNET OF ME!

1.   IDC, "The Internet of Things: Data from Embedded Systems Will Account for 10% of Digital Universe by 2010," *Digital Universe Report*, April 2014, http://www.emc.com/leadership/digital-universe/2014iview/internet-of-things.htm.

2.   Mark Cuban, "Is Search Changing?," Blog Maverick, January 18, 2014, http://blogmaverick.com/2014/01/18/is-search-changing/.

3.   Maeve Duggan and Aaron Smith, "Social Media Update 2013: 42% of Online Adults Use Multiple Social Networking Sites, but Facebook Remains the Platform of Choice," Pew Research Internet Project, December 30, 2013, http://www.pewinternet.org/2013/12/30/social-media-update-2013/.

4. Matt Clinch, "3-D Printing Market to Grow 500% in 5 Years," CNBC, April 1, 2014, http://www.cnbc.com/id/101542669.

CHAPTER 9: THE MASS CUSTOMIZATION OF ENTERTAINMENT

Epigraph. James Surowiecki, "Groupon Clipping," *New Yorker,* December 20, 2010, http://www.newyorker.com/magazine/2010/12/20/groupon-clipping.

1. Erica Ogg, "Reflecting on the DTV Transition: CEA President Gary Shapiro Talks about the 25-Year Process of Bringing the U.S. into the Digital Television Era, and Where We Go from Here," CNET, August 4, 2009, http://www.cnet.com/news/reflecting-on-the-dtv-transition/.

2. Ibid.

3. Chris Anderson, *The Long Tail,* Wired 10:10 (October 2004), http://archive.wired.com/wired/archive/12.10/tail.html.

4. Ibid.

5. "Domestic Movie Theatrical Market Summary 1995 to 2014," *The Numbers: Where Data and the Movie Business Meet,* no date, http://www.the-numbers.com/market/.

6. Lauren Coleman-Lochner and Lindsey Rupp, "Barnes & Noble Agrees to Spin Off Nook Unit As Sales Decline," *Bloomberg,* June 25, 2014, http://www.bloomberg.com/news/2014-06-25/barnes-noble-approves-nook-e-reader-spinoff-as-sales-decline.html.

7. Brent Lang, "Digital Home Entertainment to Exceed Physical by 2016, Study Finds," *Variety*, June 3, 2014, http://variety.com/2014/digital/news/digital-home-entertainment-to-exceed-physical-by-2016-study-finds-1201207708/.

8. Andrew Keen, "Web 2.0: The Second Generation of the Internet Has Arrived. It's Worse Than You Think," *Weekly Standard*, February 14, 2006, http://www.weeklystandard.com/Content/Public/Articles/000/000/006/714fjczq.asp.

9. "'Breaking Bad' Finale Soars to Series-Best 10.3 Million Viewers," *Variety*, September 30, 2013, http://variety.com/2013/tv/news/breaking-bad-finale-ratings-1200681920/.

10. Eric Schmidt and Jared Cohen, *The New Digital Age: Reshaping the Future of People, Nations, and Businesses* (London: John Murray, 2013), 27.

11. Lawrence Lessig, "It Is about Time: Getting Our Values around Copyright Right," 2009 Educause Congress, November 5, 2009.

12. Jessica Litman, "The Exclusive Right to Read," *Cardozo Arts & Entertainment Law Journal* 13 (1994): 29.

CHAPTER 10: HEALTHCARE IN A DIGITAL AGE

Epigraph. Vinod Khosla, "Technology Will Replace 80% of What Doctors Do," *Fortune*, December 4, 2012, http://fortune.com/2012/12/04/technology-will-replace-80-of-what-doctors-do/.

1. Susannah Fox and Maeve Duggan, "Health Online 2013," Pew Research Internet Project, January 15, 2013, http://www.pewinternet.org/2013/01/15/health-online-2013/.

2. Susannah Fox and Maeve Duggan, "Tracking for Health," Pew Research Internet Project, January 15, 2013, http://www.pewinternet.org/2013/01/28/tracking-for-health/.

3. "Squeezing Out the Doctor: The Role of Physicians at the Centre of Health Care Is under Pressure," *Economist*, June 2, 2012, http://www.economist.com/node/21556227.

4. Kevin Leonardi, "RNA Combination Therapy for Lung Cancer Offers Promise for Personalized Medicine: Researchers Improve Therapeutic Response in Clinically Relevant Model of Lung-Tumor Growth," MIT News, August 14, 2014, https://newsoffice.mit.edu/2014/rna-therapy-lung-cancer-personalized-medicine-0814.

5. Rui Want, et al., "Studentlife: Assessing Mental Health, Academic Performance, and Behavioral Trends of College Students Using Smartphones," *Proceedings of the 2014 ACM International Joint Conference on Pervasive and Ubiquitous Computing*, 2014.

6. "Squeezing Out the Doctor."

7. Lisa Bernard-Kuhn, "Conquering the Doctor Shortage: Expanded Roles for Nurse Practitioners and Physician Assistants Could Ease the Manpower Void, Experts Say," *USA Today*, February 8, 2014, http://www.usatoday.com/story/news/nation/2014/02/08/conquering-the-doctor-shortage/5307965/.

8. Peter Groves, Basel Kayyali, David Knott, Steve Van Kuiken, "The 'Big Data' Revolution in Health Care: Accelerating Value and Innovation," McKinsey & Company, January 2013, http://www.mckinsey.com/~/media/mckinsey/

dotcom/client_service/healthcare%20systems%20and%20services/pdfs/the_
big_data_revolution_in_healthcare.ashx.

9. Sriram Somanchi, Samrachana Adhikari, Allen Lin, Elena Eneva, and Rayid
    Ghani, "Early Code Blue Prediction Using Patient Medical Records," https://
    d277f6d674b2cfd0d2436b2145030d5d731cac78.googledrive.com/host/0B0
    TBaU3UgQ0Da3A2S2VWNTRzc1E/25.pdf.

CHAPTER 11: OF POLITICS, DATA, AND DIGITAL REVOLUTIONS

Epigraph. Charlie Rose, "Charlie Rose Talks to Wael Ghonim," *Bloomberg
Businessweek*, February 16, 2012, http://www.businessweek.com/articles/
2012-02-16/charlie-rose-talks-to-wael-ghonim.

1. Matt Peckham, "Did It Work? 'Day After' Results of the SOPA, PIPA Blackout,"
    Time, January 19, 2012, http://techland.time.com/2012/01/19/did-it-work-day-
    after-results-of-the-sopa-pipa-blackout/?xid=gonewsedit; Jenna Wortham, "With
    Twitter, Blackouts and Demonstrations, Web Flexes Its Muscle," *New York
    Times*, January 18, 2012, http://www.nytimes.com/2012/01/19/technology/
    protests-of-antipiracy-bills-unite-web.html?ref=technology&_r=1&.

2. Dan Worth, "Web Inventor Tim Berners-Lee Slams SOPA and PIPA Legisla-
    tion," V3, January 18, 2012, http://www.v3.co.uk/v3-uk/news/2139758/
    web-inventor-tim-berners-lee-slams-sopa-pipa-legislation.

3. Jonathan Weisman, "In Fight over Piracy Bills, New Economy Rises against
    Old," *New York Times*, January 18, 2012, http://www.nytimes.
    com/2012/01/19/technology/web-protests-piracy-bill-and-2-key-senators-
    change-course.html?hp.

4. Ibid.

5. Richard Verrier, "MPAA's Chris Dodd Takes Aim at SOPA Strike," *Los Angeles
    Times*, January 17, 2012, http://latimesblogs.latimes.com/entertainmentnews
    buzz/2012/01/dodd-lashes-out-at-sopa-strike.html.

6. Jenna Wortham, "Public Outcry over Antipiracy Bills Began as Grassroots
    Grumbling," *New York Times*, January 19, 2012, http://www.nytimes.
    com/2012/01/20/technology/public-outcry-over-antipiracy-bills-began-as-
    grass-roots-grumbling.html?pagewanted=1&_r=1&ref=technology&.

7. Letter from Jonathan Abrams, et al., to Members of the United States Congress,
    Los Angeles Times, no date, http://opinion.latimes.com/files/entrepreneurs-
    worried-about-pipa.pdf.

8. Gary Shapiro, "The Copyright Lobby Comeuppance," *The Hill*, December 12, 2011, http://thehill.com/blogs/congress-blog/technology/198693-the-copyright-lobby-comeuppance.

9. My account of these events relies on Wortham, "Public Outcry."

10. Tom Cheredar, "Go Daddy Loses over 37,000 Domains due to SOPA Stance," VB News, December 24, 2011, http://venturebeat.com/2011/12/24/godaddy-domain-loss/.

11. Cecilia Kang, "At CES 2012, Proposed Anti-Piracy Legislation Is Hot Topic," *Washington Post*, January 11, 2012, http://www.washingtonpost.com/business/economy/at-ces-2012-proposed-anti-piracy-legislation-is-a-hot-topic/2012/01/11/gIQADQw5rP_story.html.

12. "Time Is Running Out For SOPA Opponents Congressmen Warn At CES 2012," *Forbes*, January 11, 2012, http://www.forbes.com/sites/erik-kain/2012/01/11/time-is-running-out-for-sopa-opponents-congressmen-warn-at-ces-2012/.

13. "CES 2012: Wyden, Issa Decry SOPA; Mildly Hopeful for their OPEN Bill," Fox News, January 11, 2012, http://www.foxbusiness.com/technology/2012/01/11/ces-2012-wyden-issa-decry-sopa-mildly-hopeful-for-their-open-bill/.

14. "Media Cloud Controversy Mapper: Mapping the SOPA/PIPA Debate," Berkman Center for Internet & Society at Harvard University, http://cyber.law.harvard.edu/research/mediacloud/2013/mapping_sopa_pipa/.

15. Yochai Benkler, Hal Roberts, Robert Faris, Alicia Solow-Niederman, and Bruce Etling, Bruce, "Social Mobilization and the Networked Public Sphere: Mapping the SOPA-PIPA Debate," Berkman Center Research (Publication No. 2013–16), July 19, 2013, http://dx.doi.org/10.2139/ssrn.2295953.

16. Ibid.

17. Wortham, "With Twitter, Blackouts."

18. Amanda Terkel, "Project ORCA: Mitt Romney Campaign Plans Massive, State-of-the-Art Poll Monitoring Effort," *Huffington Post*, November 1, 2012, http://www.huffingtonpost.com/2012/11/01/project-orca-mitt-romney_n_2052861.html.

19. Michael Kranish, "ORCA, Mitt Romney's High-Tech Get-Out-the-Vote Program, Crashed on Election Day," *Boston Globe*, November 10, 2012, http://www.boston.com/news/politics/2012/president/candidates/

romney/2012/11/10/orca-mitt-romney-high-tech-get-out-the-vote-program-crashed-election-day/gfIS8VkzDcJcXCrHoV0nsI/story.html.

20. Sean Gallagher, "Inside Team Romney's Whale of an IT Meltdown," Ars Technica, November 9, 2012, http://arstechnica.com/information-technology/2012/11/inside-team-romneys-whale-of-an-it-meltdown/.

21. X Zhuo, B. Wellman, and Yu J., "Egypt: The First Internet Revolt?," *Peace Magazine*, July-September, 2011, 6–10.

22. Philip N. Howard and Muzammil M. Hussain, *Democracy's Fourth Wave? Digital Media and the Arab Spring* (Oxford University Press, 2013), 120.

23. Henry Farrell, "The Consequences of the Internet for Politics," *Annual Review* 15 (2012): 35–52.

24. Robert Brym, et al., "Social Media in the 2011 Egyptian Uprising," *British Journal of Sociology*, 65 (2014): 266–92.

25. Marc Lynch, "After Egypt: The Limits and Promise of Online Challenges to the Authoritarian Arab State," *Perspectives on Politics* 9, Issue 02 (June 2011): 301–10.

26. Patrick O'Connor, "Political Ads Take Targeting to the Next Level: Big Data Plus Detailed TV-Viewership Information Helps Reach Desired Voters for Lower Cost," *Wall Street Journal*, July 14, 2014, http://online.wsj.com/articles/political-ads-take-targeting-to-the-next-level-1405381606.

CHAPTER 12: CULTURE SHOCK

Epigraph. Julia Angwin, *Dragnet Nation: A Quest for Privacy, Security, and Freedom in a World of Relentless Surveillance* (New York: Times Books, 2014), 154.

1. Those who point to a second-century work by a Greek writer, Lucian of Samosata, stretch the definition of science fiction to a point where it's all but meaningless. In it the writer travels to the moon on a boat, having been blown to the lunar surface by a wind.

2. Anthony Lane, "Only Make Believe: 'Her,' 'The Secret Life of Walter Mitty,' and 'Saving Mr. Banks,'" *New Yorker*, December 23, 1013, http://www.newyorker.com/magazine/2013/12/23/only-make-believe-2?utm_source=tny&utm_campaign=generalsocial&utm_medium=facebook.

3. Joanna Stern, "Cellphone Users Check Phones 150x/Day and Other Internet Fun Facts," ABC News, May 29, 2013, http://abcnews.go.com/blogs/technology/2013/05/cellphone-users-check-phones-150xday-and-other-internet-fun-facts/.

4. Allison Stadd, "79% of People 18–44 Have Their Smartphones with Them 22 Hours a Day [Study]," Mediabistro, April 2, 2013, http://www.mediabistro.com/alltwitter/smartphones_b39001.

5. Susannah Fox and Lee Rainie, "The Web at 25 in the U.S.: The Overall Verdict: The Internet Has Been a Plus for Society and an Especially Good Thing for Individual Users," Pew Research Internet Project, February 7, 2014, http://www.pewinternet.org/2014/02/27/the-web-at-25-in-the-u-s/.

6. April Underwood, "Gender Targeting for Promoted Products Now Available," Twitter Advertising Blog, October 25, 2012, blog.twitter.com/2012/gender-targeting-for-promoted-products-now-available.

7. Aaron Smith, "Real Time Charitable Giving: Why Mobile Phone Users Texted Millions in Aid to Haiti Earthquake Relief and How They Got Their Friends to Do the Same," Pew Research Internet Project, January 12, 2012, http://www.pewinternet.org/files/old-media/Files/Reports/2012/Real%20Time%20Charitable%20Giving.pdf.

8. "2013 Year in Review: The Digital Giving Index: Insights and Trends on $190 Million in Donations to 40,000 Charities," Network for Good, no date, http://www1.networkforgood.org/digitalgivingindex.

9. Dana Boyd, *It's Complicated: The Social Lives of Networked Teens* (New Haven, CT: Yale University Press, 2014), 24.

10. Janine M. Zweig, et al., "Technology, Teen Dating Violence and Abuse, and Bullying," Justice Policy Center of the Urban Institute, July 2013, http://www.urban.org/uploadedpdf/412891-Technology-Teen-Dating-Violence-and-Abuse-and-Bullying.pdf.

11. See Mitch van Geel, "Bullying Associated with Suicidal Thoughts, Attempts by Children, Adults," JAMA Pediatrics Releases, March 10, 2014, http://media.jamanetwork.com/news-item/bullying-associated-with-suicidal-thoughts-attempts-by-children-adolescents/.

12. Fox and Rainie, "The Web at 25."

13. Amanda Lenhart and Maeve Duggan, "Couples, the Internet, and Social Media," Pew Research Internet Project, February 11, 2014, http://www.pewinternet.org/2014/02/11/couples-the-internet-and-social-media/.

14. Ibid.

15. Ibid.

16. Michelle Starr, "Brain-to-Brain Verbal Communication in Humans Achieved for the First Time," CNET, September 3, 2014, http://www.cnet.com/news/brain-to-brain-verbal-communication-in-humans-achieved-for-the-first-time/.

17. "Watch: Coming Early 2015" (product description), Apple, no date, http://www.apple.com/watch/features/.

18. R. M. Anderson, "With Big Data, Moneyball Will Be on Steroids," *Newsweek*, July 24, 2014, http://www.newsweek.com/2014/08/01/big-data-about-change-how-baseball-teams-evaluate-player-defense-260972.html.

19. Matthew Futterman, "Deee-fense: Baseball's Big Shift," *Wall Street Journal*, March 27, 2014.

20. Matthew Shaer, "Cage Match" *Popular Science*, August 27, 2012, http://www.popsci.com, http://www.popsci.com/science/article/2012-07/cage-match.

21. Sonali Basak, "U.S. Soccer Team Tracks Movement to Prevent Onset of Injury," Bloomberg, July 2, 2014, http://www.bloomberg.com/news/2014-07-01/u-s-soccer-team-tracks-movement-to-prevent-injury-onset.html.

22. Max Cherney, "The Technology behind the World Cup's Advanced Analytics," Motherboard, June 23, 2014, http://motherboard.vice.com/read/this-system-turns-the-beautiful-game-into-big-data.

CHAPTER 13: ECONOMICS AND BUSINESS IN A DIGITAL AGE

Epigraph. Nolan Bushnell, "What's Next, Nolan? 10 Predictions for the Next 10 Years," *SAP Business Innovation*, April 1, 2014, http://blogs.sap.com/innovation/innovation/whats-next-nolan-10-predictions-next-10-years-01247661.

1. Republican National Committee, "Petition in Support of Innovative Companies like Uber," no date, https://gop.com/support-uber-petition/.

2. Luz Lazo, "Cab Companies Unite against Uber and Other Ride-Sharing Services," *Washington Post*, August 10, 2014, http://www.washingtonpost.com/local/trafficandcommuting/cab-companies-unite-against-uber-and-other-ride-share-services/2014/08/10/11b23d52-1e3f-11e4-82f9-2cd6fa8da5c4_story.html.

3. Jaron Laniet, *Who Owns the Future?* (New York: Simon & Schuster, 2013), 2.

4. Claire Caine Miller, "Will You Lose Your Job to a Robot? Silicon Valley Is Split, " *New York Times*, August 6, 2014, http://www.nytimes.com/2014/08/07/upshot/will-you-lose-your-job-to-a-robot-silicon-valley-is-split.html?src =twr&smid=tw-upshotnyt&_r=1&abt=0002&abg=0.

5. Ibid.

6. Tyler Cowen, "Will You Lose Your Job to a Robot?," Marginal Revolution, August 6, 2014, http://marginalrevolution.com/marginalrevolution/2014/08/will-you-lose-your-job-to-a-robot.html.

7. Erik Brynjolfsson and Andrew McAfee, *The Second Machine Age: Work, Progress, and Prosperity in a Time of Brilliant Technologies* (New York: W. W. Norton & Company, 2014), 9.

8. Blaire Briody, "The Robot Reality: Service Jobs Are Next to Go," *Financial Times*, March 26, 2013, http://www.cnbc.com/id/100592545#.

9. Jill Lepore, "The Disruption Machine: What the Gospel of Innovation Gets Wrong," *New Yorker*, June 23, 2014, http://www.newyorker.com/magazine/2014/06/23/the-disruption-machine.

10. John Hagel, "The Disruption Debate—What's Missing?," Edge Perspectives with John Hagel, May 2014, http://edgeperspectives.typepad.com/edge_perspectives/2014/06/.

11. Drake Bennett, "The Innovator's New Clothes: Is Disruption a Failed Model?," *Bloomberg Business Week*, June 18, 2014, http://www.businessweek.com/articles/2014-06-18/the-innovators-new-clothes-is-disruption-a-failed-model.

12. Hagel, "The Disruption Debate."

13. Ibid.

14. See "Table 2.3.5. Personal Consumption Expenditures by Major Type of Product," U.S. Department of Commerce, Bureau of Economic Analysis, http://www.bea.gov/about/BEAciting.htm.

15. Hagel, "The Disruption Debate."

CHAPTER 14: THE PARALLEL EVOLUTION OF DIGITAL DATA, LAW, AND PUBLIC POLICY

Epigraph. Alan Turing, "Computing Machinery and Intelligence," *Mind: A Quarterly Review of Psychology and Philosophy* 59, no 236 (1950): 460.

1. Lee Rainie, et al., "Anonymity, Privacy, and Security Online," Pew Research Internet Project, September 5, 2013, http://www.pewinternet.org/2013/09/05/anonymity-privacy-and-security-online/.

2.   "Unlicensed Spectrum and the American Economy: Quantifying the Market
      Size and Diversity of Unlicensed Devices," Consumer Electronics Association,
      http://www.ce.org/CorporateSite/media/gla/CEAUnlicensedSpectrumWhitePaper-
      FINAL-052814.pdf.

CHAPTER 15: THE YEAR IS 1450...

      Epigraph. Edward O. Wilson, *Consilience: The Unity of Knowledge* (New
      York: Alfred A. Knopf, 1998), 294.
1.   Peter-Paul Verbeek, *Moralizing Technology: Understanding and Designing the
      Morality of Things* (Chicago: University of Chicago Press, 2011), 1.
2.   Ibid, 2.

Index